When the Earth Moves

When the Earth Moves

Rogue Earthquakes, Tremors, and Aftershocks

Patricia Barnes-Svarney

Thunder's Mouth Press
New York

WHEN THE EARTH MOVES:
Rogue Earthquakes, Tremors, and Aftershocks

Published by
Thunder's Mouth Press

Copyright © 2007 by Patricia Barnes-Svarney

First Printing October 2007

Thunder's Mouth Press books are available at special discounts for bulk purchases in the United States by corporations, institutions, and other organizations. For more information, please contact the Special Markets Department at the Perseus Books Group, 2300 Chestnut Street, Suite 200, Philadelphia, PA 19103, or call (800) 255-1514, or e-mail special.markets@perseusbooks.com.

All rights reserved. No part of this publication may be reproduced or transmitted in any form or by any means, electronic or mechanical, including photocopy, recording, or any information storage and retrieval system now known or to be invented, without permission in writing from the publisher, except by a reviewer who wishes to quote brief passages in connection with a review written for inclusion in a magazine, newspaper, or broadcast.

Library of Congress Cataloging-in-Publication Data is available.

ISBN-13: 978-1-56025-972-5
ISBN-10: 1-56025-972-8

Interior design by Maria E. Torres

Printed in the United States of America

To Dr. Gail Gibson and Betty Gibson—brilliant geologists and truly wonderful friends . . .

THE EARTHQUAKE

But oh! What means that ruinous
    roar? Why fail
These tottering feet? Earth to its
    centre feels
The Godhead's power and trembling
    at his touch
Through all its pillars, and in every
    pore,
Hurls to the ground, with an convolutive
    heave,
Precipitating domes, and towns, and
    towers,
The Work of ages. Crushed beneath
    the weight
Of general devastation, millions find
One common grave! e'en a widow
    left
To wail her sons! the house that
    should protect,
Entombs his master! and the faith-
    less plain,
If there be cries for help, with
    sudden yawn
Starts from beneath him. Shield
    me gracious Heaven!

—An anonymous work that appeared in the *Louisiana Gazette,* Saint Louis, March 7, 1812, a month after a round of horrendous earthquakes struck along the New Madrid fault zone

Contents

Prologue · **xi**

THE BASICS – QUAKES, SHAKES, AND AFTERSHOCKS

    Chapter 1 || When the Earth Shakes · **3**

    Chapter 2 || Necessary Faults · **21**

    Chapter 3 || Quake Parts · **33**

    Chapter 4 || The Measuring Mess · **49**

    Chapter 5 || Quirks of Quakes · **67**

QUAKES IN THE WRONG PLACES

    Chapter 6 || The New Madrid Disaster · **81**

    Chapter 7 || More Cases of Rogue Earthquakes · **105**

    Chapter 8 || Bad Effects · **147**

    Chapter 9 || Bizarre Quakes out of Nowhere · **193**

FUTURE QUAKES: WHAT'S THE RISK?

    Chapter 10 || Forecasting Earthly Motions · **207**

    Chapter 11 || What Can We Do? · **223**

    Chapter 12 || The Best Defenses Are Good Resources · **247**

Image Credits · **263**

Index · **265**

Prologue

It was just after noon, on a bright October day. I was coming down Third Street. The only objects in motion anywhere in sight in that thickly built and populous quarter were a man in a buggy behind me and a streetcar wending slowly up the cross street. Otherwise, all was solitude and a Sabbath stillness. As I turned the corner, around a frame house, there was a great rattle and jar, and it occurred to me that here was an item!—no doubt a fight in that house. Before I could turn and seek the door, there came a terrific shock; the ground seemed to roll under me in waves, interrupted by a violent joggling up and down, and there was a heavy grinding noise as of brick houses rubbing together. I fell up against the frame house and hurt my elbow. I knew what it was now . . . a third and still severer shock came, and as I reeled about on the pavement trying to keep my footing, I saw a sight! The entire front of a tall four-story brick building on Third Street sprung outward like a door and fell sprawling across the street, raising a great dust-like volume of smoke!

—Mark Twain, after a brief stint as a Confederate soldier,
headed west with his Unionist brother
to see the Wild West—including San Francisco in 1865.
This is the account of his first experience with an earthquake.

Mark Twain's observations typify San Francisco, the famous city located near the most active earthquake region in North America—and the world. Quakes occur in that part of California for the right reason: the area lies right along the infamous San Andreas Fault. But what Twain probably never realized—and most people today don't realize—is that destructive earthquakes don't always happen where logic dictates.

The Earth can truly move in all the wrong places.

Earthquakes fall into the category of natural hazards, joining the list of such often catastrophic events as hurricanes, floods, tornadoes, volcanic eruptions, and avalanches. We usually know days in advance when a hurricane or cyclone has its sights on land; tornado signatures on a weather bureau's Doppler radar can help meteorologists advise residents to run for cover; computer weather models often predict the coming of a flood-carrying weather system in time to warn people in flood-prone regions; potential volcanic eruptions can often be detected based on released gases and an increase in tremors; and even mountain avalanches can be surmised based on snowfall rates and angles of slopes.

But of all the natural disasters that strike our world, earthquakes are the sneakiest. In quake-prone areas, scientists can say that an earthquake may occur in the future, but when, and how strong, is always a matter a conjecture. They can't see it coming—and most places prone to shaking can only try to prepare for the worst.

It's even worse in places where earthquakes are uncommon. They can't see it coming, they don't know how strong the shaking will be, *and*, in most cases, they don't even realize the potential for shaking. Earthquakes in the wrong places take the word sneaky to new heights.

Quakes occur along enormous fractures in the crust, some shallow, some deep. No one has the ability to see into our world's relatively thin crust, and even if we could, there is as yet no way to collect and understand all the variables leading to a complex quake. One violent shake has so many causes—a shift here, a spreading out of a crack there—that no current computer model can keep up.

Our most notorious shakes come from earthquake zones located at specific areas—the boundaries between the irregular-shaped chunks of the Earth's crust, called tectonic, or crustal, plates. These plates—there are more than a dozen on the planet—are what scientists refer to when they discuss how our continents "drift" around the Earth over time. As each plate moves, it carries singly or a combination of ocean floor, continents, and continental shelves around the planet. For billions of years, these pieces have split and changed position like some sort of gigantic 15 puzzle. Today is no exception: we are literally walking, living, playing, and working on moving land.

And we often see the results. For instance, the December 2004 earthquake off the west coast of northern Sumatra created horrific tsunamis that smashed into coastlines around the Indian Ocean.

What about earthquakes in what I call the "wrong" places? Try the New Madrid fault zone, right next to the Mississippi River, or the fractures in the crust that run through parts of New York, central North Carolina, and New Mexico. Now add to this inventory all the other potentially destructive Earth movements in the wrong places all over the world: shaking by volcanoes and hotspots caused by massive flows of hot magma reaching the surface in strange places; collapsing rock from ground shakes that lead to landslides; or even slides occurring underwater—unseen, yet potentially deadly— along almost every continental coastline; and the list goes on.

We've only recently paid any attention to such uncommon quaking. In December 1999, scientists from all over the world presented the first global seismic hazard map as part of Global Seismic Hazard Assessment Project (GSHAP)—an earthquake hazard map developed by 500 scientists over seven years. Each year since that time, new data have been added to update the maps. Point to the deepest colors on the map—indicative of quake-prone areas—and there are no surprises. Places like Southern California, Iceland, Turkey, Taiwan, and the India-China border are major earthquake hazards. It's logical, since they're all located near plate boundaries. And there are places that don't seem to have a worry in the world—like Florida and North Dakota, the two spots in the United States that experience the fewest earthquakes each year.

But there were colorful parts of the map that caused consternation and concern: the map highlighted places away from plate boundaries that have the potential to shake to excess one day, including the Midwest, upstate New York, southern Africa, Europe, and Australia. Even New York City was not exempt.

And this map represented only earthquakes on land, not in the oceans. There is plenty of potential for marine quake zones in the wrong places—some of which can affect populated coastlines and islands, not only with shaking, but with monster tsunami waves out of nowhere. Yes, the Earth is always in motion—and we as its inhabitants are the recipients of the thrashing about. Luckily, we live in an age in which technology and knowledge can help us adapt—or at least cope—with such movements . . . at least to a point.

This book will be your guide to the actual and predictive possibilities of movements around the globe. First, we'll look at the basics of earthquakes—what scientists know and don't know, mostly based on shakings along the tectonically active borders. Next, we'll concentrate on the historical records of quakes in the "wrong places"—from the central United States all the way to the South Pole. We'll explore the myriad reasons why it's so hard to forecast or psychically predict any quake—in a quake-prone area or not. And finally, because no book can cover every fact, this book offers some resources to extend your knowledge of these natural hazards.

Who knows? You may discover that one of the potential shaky spots lies in your own backyard. Now wouldn't that be something...

THE BASICS – QUAKES, SHAKES, AND AFTERSHOCKS

CHAPTER 1

When the Earth Shakes

Not all tremors cause pain and suffering. On November 18, 1952, a slight tremor shook the town of Quincy, just northwest of Tallahassee, Florida; one record noted, "The shock interfered with writing of a parking ticket."

—Quote from media report, 1952

It was around 6:15 A.M. Pacific standard time on a warm April day in Palo Alto, California. I was enjoying a grapefruit-size orange—something my home state of New York could never produce unless the continent suddenly shifted south and moved us closer to the equator. As I wandered around the hotel room looking for a tissue in which to put the peel, I heard a deep rumble, almost as if thunder were sounding in the distance. I glanced outside and saw the sunshine—and then felt my feet shift ever so slightly beneath me. I caught myself on a nearby chair, dropping the orange in the process. For a few more seconds, it sounded as if someone were either pushing a huge breakfast cart—or driving a Harley-Davidson with a bad muffler—down the hallway outside my door. After a half minute's refection, I realized it was an earthquake—somewhat typical for this area, so close to a tectonic plate boundary. Even though the morning news announced it was "a mere magnitude 4.6 on the Richter scale," I still felt as if I had just joined a unique club, for a New Yorker—the club of those who've felt a decent shaking from an earthquake.

Little did I know I didn't need to be in California to enter the club. A few years later at my home in upstate New York, far from any of the roaming plate boundaries, another quake struck—again in the morning, around ten o'clock. I had just finished my second cup of morning tea when I heard that telltale rumble again. This time, there was just a little shaking, as if a truck were passing on a street nearby, a mere burp. I doubted my earthquake expertise. Garbage trucks were known to come by at that time, and I ascribed it to my overactive imagination. Later that morning, the radio gave the news: our area had experienced a mild shake from a quake with its epicenter at Ardsley, New York, measuring a magnitude 4.0 on the Richter scale. It wasn't my imagination; my quake "expertise" was right on target.

Another occurred on Saturday, April 20, 2002. This time I was not even aware of the shaking from the quake, which hit at 6:50 A.M. Even this moderate shaking, measuring 5.1 on the Richter scale, with the epicenter about 15 miles (24 kilometers) southwest of Plattsburgh, New York, was ignored, a "victim" of our modern lives. I had moved 10 miles (16 kilometers) away; I once again assumed it was a truck rumbling past or maybe that the landing patterns at the nearby airport had changed. But others who were more aware felt the quake—a shaking that stretched from Cleveland to Maine and Maryland, and even into New York City.

The people near Au Sable Forks in the Adirondacks are more likely to remember that morning—it was one of the most

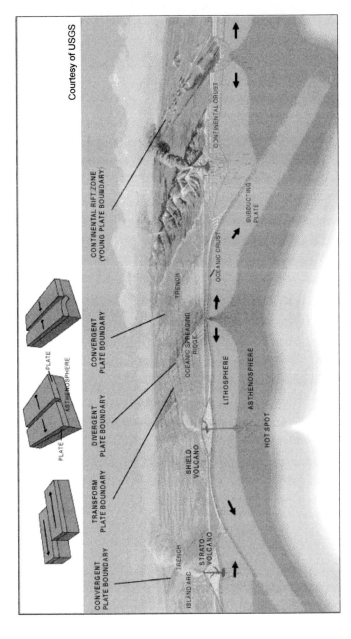

Various plate boundary types and their cross sections. *Courtesy of USGS.*

damaging earthquakes to hit the area in years, destroying enough infrastructure in six upstate counties to call in the Federal Emergency Management Agency (FEMA) for help. Adding to the destruction was another jolt: on June 25, 2002, there was a moderate aftershock in the Au Sable Forks area. More shaking; more destruction—the modus operandi in the world of earthquakes.

But wait a minute: New York is far away from areas notorious for earthquakes. How did such shaking manage to reach this part of the world? And just how many earthquakes have I felt in this region of New York—and didn't even realize it? Probably plenty, but before we delve into those wayward quakes, it's time to understand quake basics.

THE BASICS OF SHAKING
Alas, It Does Move
Earthquakes are the quintessential bad boys of the geologic world. They rarely announce their arrival, hitting without warning. In terms of landscape, they move mountains, flatten lands, and shift rivers. In terms of people—depending on its strength, location, and depth, and the material it propagates through, a quake can wipe out anything from a few homes to entire villages and cities. In fact, in the last century alone, quakes killed off at least a million people and left tens of millions homeless.

The initial quake jolt passes through the ground in overwhelming waves. To those going through the quake it feels like hours, but in reality, the shaking only lasts from seconds to a few minutes. And though most are periodically expected in certain regions of the world, the shaking waves can sometimes bounce around, causing chaos far from their source.

The biggest culprit is the Earth's thin *crust*, a layer forged during the early solar system, over 4.56 billion years ago. It is the Earth's top layer, cooled over time to create the crust—often compared to the skin over a cooled batch of homemade chocolate pudding, but not as tasty. Below the crust is the *mantle*, the seething and churning layer of liquid to semiliquid rock; and at the center, the superheated *inner and outer cores,* molten and solid, respectively.

The actual crust is so much more than we can see on the surface. Its thickness is not uniform, bulging under continents from 20 miles (30 kilometers) to about 50 miles (80 kilometers); under the oceans, it thins to about 5 miles (8 kilometers) and thickens to 10 miles (15 kilometers).

Humans are privy only to the top of the crust—the part we walk, play, drive, ski, run, and work on. This thin veneer is composed of a mix of materials that vary from sedimentary rocks (such as sandstone and shale) to igneous rocks (such as volcanic basalt) and metamorphics (such as banded gneiss)—with a bunch of fossils often thrown in for good measure. The geologic processes that give this thin crustal layer its characteristics are many: hot liquid rock called *magma* wells through cracks to the surface; the hardened rock is physically and chemically weathered into fine sediments and deposited into oceans and rivers; crustal folding crushes and changes rock by heat and pressure; animals die, and the bones that make it through being eaten or weathering are eventually covered by sediment; and plants die en masse, creating thick bogs of peat, coal, or flora fossil beds. All of this eventually turns to the rocky layers we see in eroded river valleys and outcrops, or along road cuts—our minute portholes into the crust.

Scientists have long known that the crust is neither seamless nor stationary, but is broken into *tectonic plates*—huge crustal sections that move across the planet anywhere from fractions of an inch to inches (millimeters to centimeters) per year, based on the location (thus, the official scientific term for moving plates, *plate tectonics*). Depending on what report you read, the Earth's crust is broken into thirteen to thirty-six plates; the most familiar tally includes eleven major plates, with five smaller plates completing the global jigsaw puzzle. These chunks are slamming together, sliding past each other, pushing under each other, or tearing each other apart, as if trying to vie for space on a planetary scale.

The largest crustal chunk is the Pacific plate. Its movements create havoc along the California coast, where it slips northwest past the southeastern-moving North American plate along the notorious San Andreas Fault; it moves past the panhandle of southeastern Alaska, then slides beneath the North American plate at southern Alaska and the Aleutian Islands. To the west, the Pacific plate affects Japan and Indonesia, where it subducts under the Eurasian plate. New Zealand formed as the Pacific plate rode (and continues to ride) up a slope at the edge of the Australian plate—a sliding and collision combination that created the Southern Alps.

Other plates are just as notorious: Iceland and its associated islands—including Surtsey, one of the world's youngest islands, which rose above the ocean surface in the mid-1960s—were created by the upwelling of magma along the Mid-Atlantic Ridge. This huge underwater mountain chain, which cuts through the Atlantic Ocean, is where the North American plate pulls away from the Eurasian plate. India has its own quirks: it is a "plate" in itself, a chunk of Madagascar that broke off, moved north, and ended up smashing into the Eurasian plate. The Tibetan Plateau and the Himalayas are the result of the merger—and are still rising as the Indian plate continues to push northward.

Boundary Limits

On a grand scale, the boundaries along each plate are major breaks in the crust. In total, there are about 27,000 miles (43,500 kilometers) of active plate margins. Most of this length sits on continental crust, marked by spectacular volcanoes—and often earthquakes—from those in Japan to the Andes of South America. Most of us have indirectly heard about these crustal breaks through media reports, because most earthquakes and volcanic eruptions occur along divisions between the plates. These borders are called *tectonic plate boundaries*.

What's the most important feature of these tectonic plates? Their movement can create earthquakes. When a quake occurs because of the movement along a plate boundary, it is called an *interplate earthquake;* if it occurs in the middle of a tectonic plate, it is called an *intraplate earthquake*.

The plate boundary breaks in the crust are classified as *divergent, convergent,* or *transform boundaries*—spreading or banging or sliding past one another at speeds measuring fractions of an inch (a few centimeters) per year. Divergent or spreading boundaries should be called the birthing places of islands and rock. As plates spread away from each other, hot molten magma oozes up between the chunks, creating new ocean floor. One well-known spreading center is the Mid-Atlantic Ridge, which cuts the seafloor from the Arctic Ocean through the Atlantic Ocean to beyond the southern tip of Africa—one of the Earth's longest undersea mountain ranges. This 6,214-mile (10,000-kilometer) ridge can be classified as one of the newest features on our planet, spreading an average of an inch (2.5 centimeters) per year, or 15 miles (25 kilometers) in a million years.

Even as you read this text, more land is being born along such spreading centers—not only in the Atlantic, but also along the southern Indo-Australian and Pacific plates and parts of the African plate. For example, East Africa holds several spreading centers, no doubt making it the site of the Earth's next major ocean. It has already ripped Saudi Arabia away from the rest of the African continent (forming the Red Sea). Scientists believe that the three crustal plates—the Arabian plate and the two parts of the African plate (the Nubian and the Somalian)—will separate within millions of years. Of these moving chunks of crust, it is thought that the East African rift zone will eventually form a new spreading center. This movement will also create a new ocean as the Indian Ocean floods the area, splitting the easternmost corner of Africa off to form a new (and very large) island.

Convergent plate boundaries are divided into three categories: continent-continent convergence creates huge mountain chains in response to the pushing—such as the Himalayas, the European Alps, and the Alaskan Range. As the name implies, these collisions cause the continental crust to crumple, break, and fold, but remain at the surface. This type of movement is also why a stroll in the Alps often reveals fossils high up in the mountains—the sediment that was deposited millions of years ago has been pushed up to form the mountainous terrain. Eventually, when the pushing stops, the boundary can become tectonically inactive; from there, the formed mountains no longer rise and are subject to erosion. The Appalachians of eastern North America formed in just this way eons ago—today they're a mere shadow of their former size and grandeur. The aforementioned collision of the

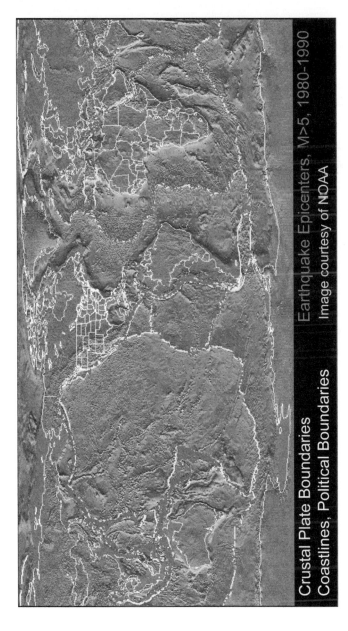

The crustal plate boundaries with coastlines and political boundaries. These dots also represent earthquakes from 1980 to 1990 greater than magnitude 5.0. *Courtesy of NOAA.*

Indian plate into the Eurasian plate 50 million years ago caused the Eurasian plate to buckle and override the Indian plate, creating the Himalayas and the Tibetan Plateau. This land, of the highest continental mountains in the world, is still growing today.

The oceanic-continental convergent plate boundaries occur when one plate dives under another plate, disappearing but still pushing the other plate. The result is usually a mountain range much like the backbone of South America, the Andes. Here lies what is known as a subduction zone—a place in which the oceanic Nazca plate is pushing into and being subducted (or diving under) the continental section of the South American plate. Many oceanic-continental convergent plates form long trenches, and this one is no exception—this is the Peru-Chile trench west of South America. These subduction zones are also regions in which active volcanoes usually reside, including those in the Andes. Another famous volcanic region from such subduction includes the Cascade Range of the Pacific Northwest.

The ocean-ocean convergent plate boundaries are also obvious: two ocean plates converging together. Like the ocean-continental borders, these plates usually subduct under one another, forming deep, long trenches. The deepest trench in the oceans formed from an ocean-ocean convergent plate: the fast-moving Pacific plate converging with the slower-moving Philippine plate. This formed the Marianas trench, a deep "ditch" that parallels the Mariana Islands. The southern end of the trench holds the Challenger Deep—so far, the deepest known spot on the Earth's surface, measuring about 35,797 feet (10,911 meters) below sea level. Here, too, the converging ocean plates often create volcanoes, most following the curve of the trench to form an island arc. The classic ones include the Philippines' Mariana Islands and the Aleutian Islands in Alaska.

Transform boundaries are crustal plate borders that slide horizontally past one another. The majority of these boundaries are found on ocean floors; a few are found on land, including the San Andreas Fault. Major characteristics include a fault that has movement in the middle, but not at the ends. Also considered to be strike-slip faults between plates, the fault motion is transformed—or changed—at the ends of the active part of the fault. They are connected to convergent and divergent boundaries. The classic case is where segments of a spreading center, such as the Mid-Atlantic Ridge, have been offset—the transform fault connects the two divergent boundaries and forms what is called a *fracture zone*. The San Andreas reigns as one of the largest transform faults, running 600 miles (965 kilometers) through California; smaller renditions are found around spreading centers. Some of these scars are long, with the offset measuring hundreds of miles to just a few meters.

Like all things in nature, there is not always a cut-and-dried way of looking at plate boundaries. In some regions, the boundaries are not well defined, due either to the deformation of the region (caused by the moving plates) or even strange and often unexplained movements and fracturing of a larger plate over time. For example, in the region between the Eurasian and African plates, there are several *microplates*—smaller fragments of a larger plate broken apart by movements over millions of years. Because of these "rogue" microplates, the interpretation of not only the extent of the plate, but the earthquake patterns and overall features of such a region, is often extremely complex.

Deepest of the Deep

Moving mantle is what most scientists believe shuffles the crustal plates. And though they have seen evidence of the mantle just by watching the hot molten magma ooze to the surface around certain volcanoes, many think that "seeing is believing," and want to see the mantle itself. Thus there have been attempts to get to the mantle by other means. One of the most recent was the third-deepest borehole undertaken by the ocean drilling vessel *JOIDES Resolution*, which reached 4,644 feet (1,416 meters) below the seafloor. The next attempt will be part of the Chikyu Hakken mission, which will use the Japanese vessel *Chikyu* ("Earth") to drill up to 23,000 feet (7,000 meters) into the ocean floor.

What is the deepest mine yet dug? Only recently, the 11,762-foot (3,585-meter; about 2.2-mile or 3.5-kilometer) East Rand mine in South Africa was the deepest; but according to reports, the Western Ultra Deep Levels mine, affectionately called "Wuddles," in West Rand, northwest South Africa, is now the winner, deepened to 3.1 miles (5 kilometers). The reasons for such "shallow" digs are obvious: heat and pressure. At around 3.1 miles (5 kilometers), temperatures reach about 158 degrees Fahrenheit (70 degrees Celsius), making it necessary for huge cooling units underground to help humans survive. Pressure of the overlying rock is tremendous, too—at 2.2 miles (3.5 kilometers), the pressure above is about 920 times the normal atmospheric pressure. When rock is removed, the pressure triples in the surrounding rock—not exactly the most pleasant of places to work.

MOVING PLATES

Driving Plates

What drives this dance of the plates? No one has yet been able to definitively describe the unseen forces that carry the tectonic plates around the planet. The mechanisms for these often catastrophic earth movements are complex, and those that cause the plates to move around the planet are still a matter of conjecture.

The idea of plate tectonics, now firmly implanted in the geologic camp, took almost a half century to find acceptance. One of the first major proponents of moving continents, German meteorologist Alfred Wegener, in 1912, had the gall to suggest that the continents moved. The jigsaw-puzzle look of the continents was observed much earlier, but it was Wegener who took the idea a step further, suggesting that the continents were like huge islands of lighter rock plowing through the seafloor's heavier rock. By "solving" the continental puzzle, he concluded that the continents had once been a single supercontinent—which he called Pangaea—surrounded by a huge ocean: a theory that shook the scientific world.

The scientific outcry was almost deafening; in fact, it was "utter damn rot," according to one disagreeable English scientist. Wegener was faced with two stumbling blocks: first, he was born at a time when scientists believed the Earth was a solid, unmoving sphere of stone spinning around the Sun, and second, he could not come up with a reasonable mechanism to explain how the continents had moved around the planet through the eons. Wegener never gave up, and over time, he would be proven right—but it would be too late. He died tragically in 1930, lost in Greenland at the age of 50—on his birthday, no less—never to be vindicated in his lifetime.

Right about the time that Wegener was lost to the world, a Scottish geologist who was influenced by Wegener came to the forefront—Arthur Holmes, a staunch supporter of the then new continental drift theory. Thanks to his work on radioactivity, geologic time, and petrogenesis, Holmes believed that there had to be something at work in the Earth's interior—something that would break up the continents, along with creating and destroying parts of the crust. He believed he had found Wegener's missing link—and proposed that the mantle possessed convection currents created by heat trapped beneath the Earth's surface. These currents welled up toward the surface, he theorized, and dragged the continents along. The light of continental drift was beginning to dawn.

> ## How Far and Where Will They Move?
>
> What are some educated guesses as to the future of crustal movement? Even now, we can see where some of the new lands and seas will be built thanks to plate movements. Though such movements will happen gradually, over the next tens of thousands to hundreds of millions of years, some areas are ripe for movement. For example, the Great Rift Valley of Africa, the Red Sea, and the Mid-Atlantic Ridge are all spreading into new positions, while the Andes and the Himalayas continue to inch up every year, growing as their respective plates continue to collide.
>
> What big chunks of crust will change? Scientists speculate that over the next 250 million years, Africa will collide with southern Europe, closing the Mediterranean; Australia will collide with Southeast Asia and China; deep subduction zones will form off the east coast of the Americas; Antarctica will move north toward India; and the Atlantic Ocean will close. Finally—and contrary to popular belief—California will not drop into the ocean, it will slide northward along the West Coast of the United States.

Slowly Proving Moving Plates
There were few champions of continental movement after Wegener and Holmes; thus, the idea of plate movement got off to a slow start. Most science books in the early half of the twentieth century chose to stick to the classic design of the Earth. As a result, acceptance of moving continents in the scientific world progressed at glacial speeds.

American scientist Harry Hess was greatly influenced by Holmes and is often credited with pushing the geologic community over the proverbial paradigm shift threshold. In the early 1960s, Hess and others mapped the seafloor around the Mid-Atlantic Ridge, trying to discern the magnetism of rock surrounding the ridge. Luckily for scientists, when magma reaches the surface and begins to solidify, certain constituents within the rock take on the magnetism present at that time, holding a small north or south pole within their makeup. Over thousands of years, as the Earth's poles change—north becoming what we call south, and vice versa—they leave their mark. The solidifying rock records those polar shifts, creating what is often called a "zebra stripe" of north and south magnetic pole readings on the ocean floor on either side of the ridge.

At first glance, it may seem as if the magma flowed over the top of the ridge, forming the stripes in magnetism. But upon closer examination, Hess deduced that the changes in north-south magnetism were the result of the movement of the ocean floor—a slow shift of the ground literally pushing away from the ridge called *seafloor spreading*. Thanks to Hess's discovery, evidence of continental drift and tectonic plate movement was beginning to come together—albeit slowly. By the 1970s, scientific textbooks and professional papers embraced the fact that the continents do move—and that the continental and oceanic crusts shift and "float" above the mantle.

Interpreting Zebra Stripes

No one knows why the north and south magnetic poles reverse every so many thousands of years. The evidence of these changes lies along the seafloor—and one in particular is in the North Atlantic Ocean near Iceland—on either side of the spreading Mid-Atlantic Ridge. Here, the rate of plate movement is about 1 inch (2.5 centimeters) per year, or about 16 miles (25 kilometers) every million years. What are some past—and current—reversal periods exhibited in this region? Representing the present time to 780,000 years ago, is the Brunhes Normal; the Matuyama Reverse was between 780,000 and 2,590,000 years ago; between 2,590,000 and 3,500,000 years ago, the Gauss Normal (a time similar to our north and south poles) occurred; and the Gilbert Reverse ran from 3,500,000 to 5,300,000 years ago.

Modern Moves

Fast-forward to the present: we definitely know that tectonic plates move. Modern observation of these crustal chunks—thanks to technology such as global positioning satellites and electronic-laser equipment—has helped us realize how complex these movements have been and continue to be. We know that the average rates of plates separating, moving, shifting, and even spinning can range widely, and it's been like this for millions (if not billions) of years. Today, on average, the Arctic ridge has the slowest rate, separating around an inch (2.5 centimeters) per year; the fastest rate, at more than 6 inches (15 centimeters) per year, is at part of the East Pacific rise near New Zealand.

Still, though plate movement is generally agreed upon, unraveling the complexity behind the moving plates has not been easy. Even today, there are more theories to explain how these earthly chunks move than there are crustal plates. The leading proposal—similar to Arthur Holmes's—is the *mantle convection theory*, in which the mobile rock beneath the rigid crustal plates is responsible for our restless Earth.

This theory states that the mantle's molten rock rises up from the intense heat near the Earth's outer core, cools as it reaches just under the crust, and sinks back toward the superheated core again. This creates a convection cell, much like the rise and fall of water boiling on a stove—or, the comparison many geologists prefer, the motions exhibited by a lava lamp. Whatever the analogy, something has to keep the conveyor belt of plates moving, shifting, and wiggling—and convection seems to fit the bill. The "floating" crustal chunks—made of mostly crust and some upper-mantle material—on the upper mantle's semisolid rock move slowly across the planet over millions of years.

Not all scientists agree that the mantle convection theory completely supplies the reason for shifting crustal plates. Some believe that seafloor spreading is the most important mechanism in plate tectonics—that a self-explanatory phenomenon called "ridge push" propels and maintains the moving plates. Others believe that subduction is a much more important process in moving plates. At convergent boundary zones, the gravity-controlled sinking of the denser, cooler ocean plate into the subduction zone is called "slab pull." In other words, no one knows what is more important in driving plate tectonics—the opening of a ridge spewing magma, the yanking of a plate underneath another, or both.

And of course, one mystery leads to another: just what is the origin of the heat governing the mantle's movements? Is it the radioactive decay at the Earth's core and residual heat—or could it be the gravitational energy left over from the formation of the planet 4.56 billion years ago?

No one quite knows which force wins out in the race to move crustal slabs across the Earth—or even what method of heating up the cores and mantle outscores the other. Presently, none of these ideas completely explains the movement of the plates. Scientists dream about drilling into the mantle at the thinner spots in the oceans—but the dreams fade when they look at the immense drilling distances and the inaccessible terrain. And because technology has a long way to go before they can reach that deep into the crust to determine the answer, these theories remain just educated guesses.

Triple Junctions

As with everything in nature, there is an overall pattern; but occasionally, there is a glitch in the system. Triple junctions are just such a glitch, and, as the name implies, are spots in which three crustal plates come together. One of the most infamous is in East Africa—a place in which the actively splitting the African plate and the Arabian plate meet in a triple junction, where the Red Sea meets with the Gulf of Aden.

Even more daunting are the three major triple junctions associated with one feature: the East Pacific rise. The Galápagos Islands are located above an alleged hotspot on the Galápagos ridge, a spreading center that extends east from a triple junction with the Pacific rise. (For more about hotspots, see chapter 8.) This large triple junction is rare in our crust—it's literally a junction that separates the Pacific, Cocos, and Nazca plates with a microplate, aptly called the Galápagos microplate. The subsurface movement of liquid rock affects the microplate in a dizzying way—causing the small chunk of crust to spin clockwise between the surrounding larger plates.

South of Easter Island, the East Pacific rise splits at another triple junction. It does this with another rise called the Chile rise; the movement is to the east, pushing the crust under the coast of southern Chile, creating a subduction zone. Finally, the East Pacific rise extends to the south, where it is called the Pacific Antarctic ridge, merging with the Mid-Indian ridge at a triple junction south of New Zealand. In fact, this southern extension is also one of the fastest-moving spreading centers on Earth.

CHAPTER 2

Necessary Faults

And the seventh angel poured out his vial into the air . . .
And there was a great Earthquake.
—*Revelations 16:17–18*

When one discusses moving mountains, one probably isn't thinking about the small island of Jamaica just south of Cuba. Back on June 7, 1692, the Jamaican town of Port Royal was bustling—frequented by pirates, prostitutes, and people seeking to procure items illegally. On that hot, sultry morning around 11:40 local time, the port—and the spit of land it was on—was shaken by three major quakes. The first quake was moderate; the last, horrific—probably hovering around magnitude 6.0 on the Richter scale. There were reports that the ground didn't shake, but moved like waves on the ocean. The unstable sands overlying limestone on which Port Royal sat caused massive damage—less to older, more flexible wood houses and buildings, more to the newer stone and brick. As water saturated the sands, people were caught in the mucky mix. Those unfortunate enough to be on land were dragged into Kingston Bay, the city literally sliding into the waters—with the far end of the town coming to rest in 50 feet (15 meters) of water.

And it didn't end there: a (very) short time later, tsunami waves caused by the earthquake flooded the leftover city streets and buildings, killing hundreds more—including many who were struggling to get out of the mucky saturated sands. Overall, the quakes took about 2,000 lives and wrecked 1,800 houses in the city alone, in only a few minutes; the tsunami wave caused even more damage. Help was hard to find, as the region around Port Royal was cut off from the rest of the long Palisade spit; looting followed almost immediately. The few survivors rebuilt on what was left of the acreage (though in 1704, the unlucky Port Royal experienced a fire that destroyed all the buildings except for the forts). For the next hundred years, visitors could see the tops of the town's former taller homes in the murky water just offshore—a reminder of the ramifications of a moving Earth.

TRUE FAULTS
Everyone's Faults

The earthquake at Port Royal was only one out of the no doubt millions (or billions) of major quakes that have occurred over the planet's 4.56-billion-year history—or one of the many hundreds since recorded time. Plates move; the planet shakes. It seems like a simple formula of action and reaction, and, in fact, the shaking from earthquakes is caused by simple physics: the tectonic plates move, creating stress and strain on the crust. All this movement causes splits not only at the plate edges, but out from the boundaries, like spreading tributaries of a river. And all these "cracks," called *faults* (also *fault lines* or *rupture surfaces*), are places ripe for quakes.

We really shouldn't call faults "cracks in the rock," as they are not typical cracks—like in a cement sidewalk. Faults in the Earth's crust follow preset conditions in a rock layer—long weak spots in the

An offset fence in California after the great San Francisco earthquake of 1906. *Courtesy of USGS.*

rock mass that cause the layers to move in relation to one another. It may be a layer of sedimentary rock or a weak spot in a wall of volcanic basalt. Whatever the specific condition, when a plate moves, a split travels along the weakest spots in the rock, which, in turn, dictates where and how a fault forms. Small versions of faults—and the resulting rock cracks, cuts, and folds—can be seen along many of the highways around the world in the form of *outcrops*. As the word implies, an outcrop is an exposure (the outward appearance of a "crop" of rock) seen in cross section. Geologists call these exposures "natural laboratories," especially when the outcrop reveals part of our crust usually covered by a thick layer of topsoil.

One of the first to explain the generation of quakes along a fault was American seismologist Henry Fielding Reid, who in 1911 studied the horrific 1906 San Francisco quake. He determined that earthquakes occurred at the crustal boundaries and in tectonic (volcanically active) areas. These ideas led him to propose the *elastic rebound theory*, which states that during a quake, the rocks under strain suddenly break, creating a fracture along a fault. When the fault slips, its crustal movement causes a small to large vibration—or an earthquake.

Reid's explanation has held up—with a bit more technical detail added—for almost a century. Scientists often use the San Andreas Fault as a prime example of the elastic rebound theory: as the North American and Pacific plates scrape by one another, no crust is being created or destroyed. But there usually is a reaction—stress and strain continue to build until a sudden release occurs. This "jump" along either side of the fault triggers an earthquake—from a slight tremor to a major quake. The mechanism is usually the same; it's the amount of stress and strain that carry weight when it comes to earthquake magnitude.

You can try a simple experiment to simulate such movements. Slightly dampen one of your hands and run it over a smooth countertop. Notice how your hand grabs (the friction holding it to the counter) and releases (the movement of your hand overcoming the friction, releasing energy) as it skims across the surface. Each one of those "jerks" is similar to how two chunks of the

Oblique aerial view of the eruption on May 18, 1980, of Mount Saint Helens in Washington State, which sent volcanic ash, steam, water, and debris to a height of 60,000 feet. The mountain lost 1,300 feet of altitude and about two-thirds of a cubic mile of material. Note the material streaming downward from the center of the plume and the formation and movement of pyroclastic flows down the left flank of the volcano. *Courtesy of NOAA.*

crust move past one another, become caught, and jump, creating a quake. In the case of rocks, the stress builds, the rock suddenly slips, and energy is released in waves that travel through rock—often causing people, places, and things to shake at the surface.

Each type of plate boundary holds its own earthquake story. The sudden compressions, slips, dips, and separations that create a quake occur seemingly at random, but in reality, it is done to a dance that makes sense on a restless, shifting planet. One of the reasons we know about the seemingly constant jostling of the crust is by the shorter, jerky shifts felt every year—mostly as minor earthquakes. Each plate moves slightly, adjusting itself as the other plates move across the globe, transferring their energy to the rocks. The natural elasticity of rocks can take most of the stress and strain, but pressure can build up, causing a weak section in the crust. And it's the failure and release of energy at these weak sections that we call earthquakes.

Scientists categorize faults into three types. A dip-slip fault includes two of these faults: normal and reverse. A *normal* fault is one that relieves tension stresses along the fault, resulting in two pieces of the crust moving apart, such as those normal faults that make up the mountains and valleys of the Western United States Basin and Range province. *Reverse* (or thrust) faults cause earthquakes that occur most often at shallow depths of 0 to 32 miles (0 to 50 kilometers) as two pieces of crust move together (it looks like a normal fault in reverse, thus the name). They are notorious around areas of compression—especially subduction zones, such as the one found off the coast of Japan. And finally, the *strike-slip* fault takes care of the lateral stresses as two crustal pieces slide by one another—the classic example being the San Andreas Fault along the California coast. It is rare that these faults exist alone;

they are usually a combination of several, such as a normal or reverse fault along a strike-slip fault.

> ### Earthquake Winner
>
> Many places on Earth experience frequent quakes, but the winner has remained the same for centuries: the Pacific Ocean's Ring of Fire. The Pacific Ocean experiences around 90 percent of the world's earthquakes, including 81 percent of the largest quakes, along with the most tsunamis and volcanic eruptions. One reason is the vastness of this ocean—along with the plate boundaries that "ring" the huge Pacific plate.
>
> In reality, the ring is actually a 24,856-mile (40,000-kilometer) horseshoe—a chain made up of ocean trenches, island arcs, and volcanic mountain ranges also called the circum-Pacific belt or the circum-Pacific seismic belt. To the east are the Nazca and Cocos plates subducting beneath the South American plate; at the northern end, the Pacific plate slides under the Aleutian Islands arc, causing Alaska to rumble frequently with major quakes. The southern extension is much more complex, mixing with less spectacular tectonic plates that are colliding with the Pacific plate—for example, the smaller plates around New Zealand, the Philippines, and Tonga.
>
> Coming in a close second in terms of seismic activity is the Alpide belt. Around 17 percent of the world's largest quakes occur in this belt, which extends from Java through Sumatra, to the Himalayas, over to the Mediterranean, and into the Atlantic Ocean. And coming in at third in seismic activity is the Mid-Atlantic Ridge, the long chain of mountains that cuts through the Atlantic Ocean floor.

Faults in the Middle

Though quakes often occur along thrust faults in subduction zones, there's an even scarier threat that can cause shaking from greater depths. These are called intraslab earthquakes, and they occur within the descending oceanic crust at depths of 32 to 186 miles (50 to 300 kilometers) beneath the surface. The shaking is caused by a different process than the thrust-fault quakes—forming as oceanic crust dives deep beneath the edge of continental crust.

Scientists are particularly wary of intraslab quakes. Though they occur near a plate boundary, intraslab earthquakes can almost be called "quakes in the wrong places." One crustal slab dives under another; the action is similar to shoving a wedge under a door. The resulting earthquake occurs deep, right where the wedge point slips under the plate. There is no way to truly determine where that "wedge point" is located—and thus, no one can truly determine how far into the interior of the plate such a quake will occur. The only saving grace is that because the quake occurs at such great depths, minor to moderate seismic waves often dissipate before they reach the surface.

This was not the case in 1949 in Washington State. The most damaging earthquake in the Pacific Northwest was a major intraslab earthquake under the city of Olympia. It measured a magnitude 7.1 on the Richter scale and caused over $100 million in damage. Scientists realize that large intraslab quakes occur quite frequently. And they're worried that places like the Pacific Northwest—and cities such as Vancouver, Seattle, and Tacoma—may be at an even higher risk for a major intraslab quake than for a quake due to a nearby offshore trench.

Just why these quakes occur is still a mystery. Scientists are particularly baffled because it doesn't seem as if such faults ought to be able to move, because of the weight of the overlying rock. One theory is that intraslab quakes shake because the intense heat and pressure in the subduction zone transforms the

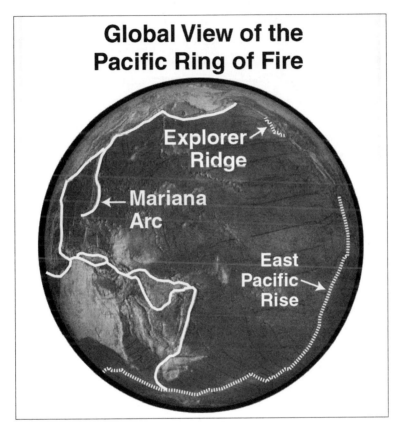

The Pacific Ocean's Ring of Fire. *Courtesy of NOAA.*

descending oceanic crust into denser rock. When this occurs, the changes in the rock can literally squeeze water from the minerals. This water, in turn, essentially lubricates the fault—the "pumped-up" water pressure causing the fault to slip and create a quake. This theory was recently tested in two Japanese subduction zones based on the earthquake activity and heat given off—one called a "warm subduction zone" in southwest Japan and the other a "cold subduction zone" beneath northeast Japan. The results showed that under the warm zone, water is freed at shallower depths and triggers only shallow intraslab quakes; in the cold zone, water is released at much greater depths and quakes are much more violent.

Other Bad Faults

It may sound like something related to NASCAR, but in geology, *relay ramps* are zones that appear as slopes—or ramps—that bridge two overlapping normal faults as a way to accommodate a crustal plate's movements. There are several famous ones in the world, including the East African rift, the Gulf of Suez, and in Italy, at the site of the famous 1688 Sannio earthquake (which measured 7.1, with its epicenter in the southern Apennine Mountains about 30 miles [48 kilometers] northeast of Naples).

There is another, seldom-mentioned type of fault that has the potential to cause a great deal of damage—especially in out-of-the-way places. It is called a *blind thrust fault*, and is a seismic fault that exists miles underground and shows no revealing traces on the surface. These faults often have few quakes associated with them, and even during and after an associated quake, the fault still shows no trace of it. Scientists know they exist all over the world. The biggest problems with these faults are that no one knows how deep they are, how quickly or slowly they slip, or how active they may be. Some of the ones that cause consternation—even though the faults are deep underground—include those around Los Angeles Basin, an area in which scientists have detected the Elysian Park and Puente Hills blind thrust faults. Another, more famous, earthquake occurred along one such fault about 10 miles (15 kilometers) deep: the 1994 Northridge quake, with a magnitude of 6.7; it killed 61 people and caused more than $20 billion in damage.

But don't be fooled into thinking that blind faults occur only near tectonic plate boundaries. Others have been discovered, like the one in the Lake County region of Tennessee—a pivotal fault line that isn't even near a plate boundary. This one (in the infamous New Madrid fault zone) was found when researchers dug long, deep trenches in the Mississippi River floodplain in the Reelfoot Lake region. The recently deformed, just-below-the-surface sediments were indicative of a blind thrust fault, as the layers absorb the upward motion of the energy released by ruptured faults. By measuring the "kinks" in the banded sediments, the researchers determined that the slip rate of the Reelfoot thrust fault was about a quarter of an inch (0.6 centimeters) per year in the past 2,300 years—equaling around 45 feet (14 meters). For more about the New Madrid fault zone, see chapter 6.

CHAPTER 3

Quake Parts

Some subterranean monster, roused from sleep,
Goes plunging through the caverns of the deep,
Stamping in fur on the earth's thin floor,
Making the vault reverberate with his roar.
—*Alex H. Sutherland, "The Earthquake," 1935*

It is often said that "the pen is mightier than the sword," but a geologist friend once witnessed that "the word is mightier than the ground." During a long, drawn-out geology conference, someone was talking about the propagation of earthquake waves when the floor started to ripple. As everyone at the conference watched, the floor heaved and the tiles seemed to move as the shaking of a moderate quake ran through the building. Everyone jumped up and someone shouted that they should run out of the building. In the meantime, the lecturer looked stunned. The resulting quake measured around 5.4 on the Richter scale, leaving a few tiles broken on the floor, tumbling some bricks from local chimneys, and cracking some buildings.

This sounds like something that could happen in California,

Armenia, or even Italy. But it didn't—it happened in a "wrong place," in central North Carolina. And rumor has it that it was a long time before the scientist-lecturer would give a talk about earthquakes again.

SEPARATE WAVES
Interpreting Quake Waves

As we've seen, sudden movements along faults are the main cause of earthquakes. Along their length, friction holds the plates together. Years to centuries to eons, the boundaries can hold together, until one day, the stress and strain becomes too much. When this type of fault boundary suddenly bursts—or the stress as the plate tries to hold itself in one spot becomes too much—an earthquake can occur, shaking the local area or regions hundreds of miles away. And it doesn't take much. One study of more than twenty faults showed that earthquakes can be triggered by as little as one-eighth the pressure required to inflate a typical car tire.

Ground movement from earthquakes can cause many failures. The faults can break the ground, creating rills and gullies in soil and even rock. The Earth may shift, moving structures in different directions depending on the fault, the classic example being the twisting of once-straight railroad tracks during a quake. Material along steep, unstable slopes (or those with a covering of loose soil or rock) may slide, especially if the quake occurs with excessive rains or snowmelt. Tall ridges may crack under the strain, splitting open along their crests. Areas in valley regions, such as those with overlying glacial till or along shallow bay areas (in which lagoons are filled with sediment buildup), can amplify the ground's vibrations.

The rupture surface does not slip at once; the shaking begins at a point on the fault plane called the *hypocenter*, located from a few to tens of miles (tens of kilometers) below the surface. In minor and major quakes reported in the media, the scientists interviewed often refer to the *epicenter*—the place on the surface where the first rocks slipped as the quake hit—in other words, directly above the hypocenter. For example, the epicenter of the Loma Prieta earthquake on October 17, 1989, occurred, logically, in Loma Prieta, California, the slip originating between 5.5 and 10 miles (9 and 16 kilometers) below the surface. (The name of a quake is most often based on its epicenter—even though the extent of the quake, or the fault's rupture surface, can be miles long and wide.)

Riding the Waves

From the sudden shift in rock, shock waves travel out in all directions; the effect is like that of dropping a rock in a calm pond. And though we just hear about a "magnitude" when news breaks about an earthquake, the shaking entails a great deal more: motions of acceleration (how fast the ground is changing), velocity (how fast the ground is moving), frequency (the vibration of the seismic waves differ in frequency), and duration (how long the quake lasts).

Body waves are the first to radiate from the initial shaking and are broken down into primary and secondary (P- and S-) waves. P-waves, as the name implies, are the first waves to head out. They can easily move through solid rock and molten materials. Also called compressional waves, they are the fastest waves that alternately compress and expand material in the same direction in which they are traveling. Next to arrive are the S-waves—the most damaging waves from an earthquake. They are also called shear waves and are slower than the P-waves, shaking the ground in an up-and-down, along with back-and-forth, motion, perpendicular to the direction in which they are traveling. By knowing how fast both the P- and S-waves travel through the Earth as recorded by at least three seismic stations, scientists can calculate the time and location of an earthquake within minutes. Determining the shape and location of the entire rupture from the seismic data takes more time—usually up to a few hours.

There are also two less-talked-about waves that fall behind both the P- and S-waves—Rayleigh and Love. These are not body waves but surface waves, and most often occur after the P- and S-waves have struck (though it is rare, they have been reported to be faster than S-waves). Both of these waves are generally stronger than the P- or S-waves, too, but their effects are confined to the surface. Rayleigh waves, named by British geophysicist John William Strutt, Lord Rayleigh, roll along the ground like waves in the ocean. As the Rayleigh wave moves continuously forward, the individual particles move vertically in an elliptical path. You can duplicate the same motion in a pond. If you toss in a stick, it will stay in one spot; toss in a stone, and though the waves will cause the stick to bob up and down, the stick still stays in one place. The second surface waves were named after British mathematician and geophysicist A. E. H. Love, who predicted the waves mathematically in 1911. They are purely transverse motion waves—or move perpendicular (side to side) to the propagation direction and parallel to the surface.

Though not a hard-and-fast rule, oftentimes the longer the fault, the bigger and longer lasting the quake. For example, the 9.2 March 27, 1964, Alaskan quake had a rupture length of 621 miles (1,000 kilometers) and lasted 420 seconds; the Denali,

After the Guatemalan earthquake of 1976, the Motagua fault showed up on the surface of this soccer field. Such "mole tracks" form best in brittle soils. *Courtesy of USGS.*

Alaska, 7.9 quake on November 3, 2002, ruptured about 186 miles (300 kilometers) and lasted 90 seconds; the Loma Prieta quake mentioned above had a magnitude of 7.0, ruptured about 25 miles (40 kilometers), and lasted 7 seconds; the 1906 San Francisco quake at magnitude 7.7 caused a 250-mile (400-kilometer) rupture and lasted 110 seconds; and the New Madrid, Missouri quakes, in 1812 alone, and measuring around magnitude 7.2 to 7.8, ruptured between 25 and 62 miles (40 to 100 kilometers) of the fault and lasted from 13 to 30 seconds.

After the initial shaking begins, it keeps spreading until something stops it. What stops it? Does the fault run out? Does the rock somehow change, stopping the fault from ripping open any further? In fact, no one actually knows why or how quakes cease.

What determines how a quake will feel to those of us on the surface? The three major factors are distance, magnitude, and local geologic conditions. In deep earthquakes, distance becomes our friend; the seismic waves diminish in intensity as they travel through the ground to the surface. The high-frequency waves from the quake dissipate rapidly with distance, while the low-frequency waves dissipate less rapidly—similar to how with distance you can hear a low-pitched noise better than a high-pitched noise. Thus, the closer to the quake, the stronger the jolt and the "faster" the up-and-down motions, while farther away, the higher frequencies will have died out, leaving only a rolling motion.

Take the magnitude 7.8 quake on January 9, 1857, in Fort Tejon, California; it lasted 130 seconds and ruptured about 220 miles (360 kilometers) of the fault. Traveling at 2 miles (3 kilometers) per second, it took just over two minutes for the length of the fault to rupture. It may sound frightening at first, but the origin of the quake was 250 miles (400 kilometers) underground, far enough away for the seismic waves to dissipate. But not completely—people and buildings just above the rupture still shook hard for about 10 to 15 seconds—though not for the entire two minutes.

One caveat, before those of you in the East start feeling better: remember, it's not only the magnitude and distance that affect the ground you stand on during such a quake—it's also the rock and soil. In the eastern part of the United States, which has different geologic conditions than the West, your quake experiences would be quite different. The same magnitude of earthquake in the West dies off with distance quickly; in the older, more rigid crust of the eastern United States, the soils and rock actually amplify the shaking produced by a quake. You will shake at greater distances, and sometimes more, in the East.

Such differences do not mean cities in the West withstand quakes better than those in the East. In certain spots, humans have managed to "change" the soil structure, making quakes react differently. For example, even though the 1989 Loma Prieta earthquake in California occurred 60 miles (100 kilometers) from San Francisco, those spots in the Bay Area sitting on landfills or soft, loose soils were shaken more intensely. In fact, the elevated Nimitz freeway collapsed because it stood on loose soils—with the ground motion more than ten times stronger than at neighboring spots sitting on rock. This is a good lesson for those living in the Los Angeles Basin, where thick, loose sediment often reaches 6 miles (10 kilometers) thick. A shaking here can be up to five or more times greater than a shaking in the nearby mountains surrounding the city.

Loose Rock; Liquid Rock

An earthquake's overall effect depends on the underlying material—whether its waves propagate through hard rock or soft sediment. And you don't want to have soft sediment in a quake zone. Take for example Mexico City, Mexico, a major metropolitan area with a population that exceeds 20 million. It stands at least 200 miles (322 kilometers) away from the place in which the Cocos tectonic plate subducts under the North American plate. When an earthquake occurs along this border, it's felt in Mexico City—and a large quake may one day tear the city apart.

Why? Several hundred years ago, the conquistadors took over the region and decided the local lake stood in the way of building a magnificent city. So they filled in the lake and built on top—resulting in Mexico City, a city that has grown to be one of the largest in Central America. But the filled-in lake bed contains clay and water; when a quake occurs around the area, the loose material acts like a seismic amplifier, causing the shaking to continue for much longer than it would if it had been built on solid rock. One such quake, in the 1970s, caused the collapse of buildings 6 to 15 stories high in the city. When the scientists looked at the quake data, they realized that the shaking through the sediment caused a resonance, or frequency, that made the clay layer vibrate. This resonance was also "in tune" with the 6- to 15-story buildings, causing them to fall apart and collapse. Since that quake, no structures built on the lake-sediment areas can be higher than 6 stories.

Quakes can also change the consistency of rock, with one of the most devastating effects called *liquefaction*—when the ground literally turns to a slimy, muddy slurry that carries debris, rips away foundations, and causes chaos. Under certain conditions, when soil and sand are exposed to moderate to high earthquakes, the materials behave like a dense fluid rather than a wet solid mass. Low-lying areas in which the water table is high are particularly vulnerable, as are areas that have excessive amounts of water runoff, which adds more water to overlying soils. When liquefaction occurs, it can cause buried tanks and pipelines to literally shake to the surface. It's the same effect as shaking a mixture of rock and soil—the larger rocks rise to the surface. At the same time, heavier objects on the top of the ground sink as liquefaction causes the soil to lose its weight-bearing strength.

Surface Ruptures

Earthquakes don't just shake the ground—very large quakes can produce a rupture, or a ground displacement, that runs for hundreds, if not thousands, of miles. Such a rupture occurred during the great Sumatran quake in December 2004, in which the length of the crustal rip extended about 746 miles (1,200 kilometers) in a gently curving line running north-south, roughly following one of the plate boundaries in the region.

Earthquakes often begin at a certain part of the rupture surface. The energy then propagates like a spreading crack in a cement wall, usually following weaker points in the rock—and sometimes a previous rupture. One of the most famous land ruptures is found at Ethiopia's East African rift zone. In fact, in 2004, scientists studying the region noted that several ruptures were found in the desert—a 20-foot- (6-meter-) wide fissure in the rift zone that spewed out ash after a set of quakes hit the region.

This site may be the beginning of an ocean—an area in which the crust is spreading apart to form an inland sea. The newest opening measures about 12 miles (20 kilometers) wide and 37 miles (60 kilometers) long, and is fractured by a series of 1.8-million-year-old faults. It is part of the Gulf of Aden triple junction (also called the Afar triple junction; for more about triple junctions, see chapter 1), a spot in which the Arabian and African plates come together—and where Africa is also splitting apart. No one quite knows why the area is coming apart—the region's tectonic forces are too weak to cause such a crack. But some scientists believe that hot magma from the mantle is welling up under the southern part of the zone, causing the rifting. And don't hold your breath to see an ocean anytime soon. Because the pieces of crust are pulling away at about a quarter of an inch (0.6 centimeters) per year, it will take millions of years—at least—to split apart.

In the oceans, large ruptures disturb the ocean surface, often creating a tsunami, or seismic sea wave—but not always. (For more about tsunamis, see chapter 8.) Sometimes the displacement of water is enough to shift the water, creating destructive waves that reach the shallow waters of a coastline. Other times, the displacement is large, but the waves—because of the location of the quake, the contours of nearby land and ocean floor, or other factors—dissipate.

Ruptures and the Movies

We've all seen the movie(s): the hero and heroine are running toward a safe haven as the earthquake continues to rumble and rock for at *least* five minutes. As they reach the top of a precipice, the evil sheriff is after them, along with the wicked banker, deceitful construction worker, and insane housewife. As the good guys run, they take a peek behind them: the ground opens up, swallowing all the various antagonists, saving the hero and heroine. They embrace, the quaking stops, and all is right with the world.

Can the Earth really swallow people up during a major quake? Not likely. It's true that cracks and fissures are commonly caused by quakes, but the notion of being "swallowed up" by the ground is just plain wrong. Yes, you could slip or trip over a surface crack caused by an earthquake, or even witness the uplifting of the land during major-magnitude quakes. But the ground rarely opens and closes, grabbing townspeople—no matter how wicked they might be.

And even animals are exempt. One favorite myth about the 1906 San Francisco earthquake was that the Earth opened up, swallowed a cow, and closed up again. Interestingly enough, this story was reported by a geologist who visited the farm and witnessed the cow's legs sticking up from the ground. Fifty years later, the truth was told: a ranch hand admitted that he had found the dead cow, tipped it into an earthquake-associated fissure, filled it in with dirt—and started the rumor.

QUAKE SHOCKS
Foreshocks and Aftershocks

In the majority of earthquakes, the major quake is not all the crust has to offer. There are the before and after quakes—aptly called *foreshocks* and *aftershocks*. Though some scientists debate the idea, others believe foreshocks—smaller quakes before a major quake—exist. They often use foreshocks to predict larger quakes, but it's not that easy. In fact, such shakings are usually "verified" in hindsight, not before an actual large quake. After all, how can you tell if the initial shaking is actually the primary quake and not just a foreshock?

One of the most cited foreshocks was a magnitude 7.9 foreshock in Chile. It occurred before the great quake of 1960, which measured magnitude 9.5—thought to be the largest quake in recorded history. Another popular foreshock citation is the 1906 quake that tore apart San Francisco. At about 5:12 A.M. local time, a foreshock occurred with enough force to be felt throughout the city. About 20 to 25 seconds later, the great 7.7 quake hit, with the epicenter near San Francisco.

A magnitude 6.8 earthquake on October 14, 1968, in Meckering, Australia, caused about $2.2 million in damages, along with this 19-mile (30-kilometer) surface rupture. *Courtesy of NOAA.*

There are places where foreshock prediction seems to work. One in particular is found at the East Pacific rise—the long north-south welt of seafloor spreading under the eastern Pacific Ocean that travels from near Antarctica north to the northern end of the Gulf of California, into the Salton Sea basin. This transform fault zone shows foreshock activity before a major seismic event. (In fact, the region's foreshock activity outweighs its aftershock events—especially when compared to nearby continental strike-slip fault activity.)

If the strain within the Earth is really bad, the initial quake can leave a disquieting aftertaste—in the form of aftershocks. As the name implies, aftershocks are generated in generally the same region as the initial main shock, occurring within predictable bounds of space, time, and magnitude. They can be the most destructive parts of earthquakes, taking structures that are already on the verge of collapse from the initial quake and pushing them over the breaking threshold.

Physically speaking, when compared to earthquakes, individual aftershocks are not that much different. They, too, include rupture processes and ground motions that are indistinguishable from the initial quake, and oftentimes they can have magnitudes just as large as the main shock. In fact, aftershocks differ from other earthquakes only in that we expect them. For this reason, in 1997, geologists Susan Hough and Lucy Jones proposed in an article in *EOS* (published by the American Geophysical Union) that the phrase "just an aftershock" be banned from the English language.

How long do aftershocks last? There are a plethora of geologic "laws" that try to explain aftershocks. One is called the modified *Omori's law,* based on work published by Fusakichi Omori in 1894. It states that the frequency of aftershocks decreases roughly by the reciprocal of time after the initial quake. (Omori also developed his own seismic intensity scale; for more about intensity scales, see chapter 4). Then there is *Bath's law,* named after Swedish seismologist Markus Bath. This rule states that any initial shock has an aftershock averaging about a magnitude 1.2 smaller than the main shock—a rule that has always been controversial, with scientists debating whether the law has a physical basis or is simply a result of statistics using aftershock data. Then there is the *Gutenberg-Richter relation*—named after American seismologist Charles F. Richter and German scientist Beno Gutenberg, both of whom also developed the Richter scale—that measures earthquake magnitude (for more about earthquake scales, see chapter 4). This relation states that as you go down one magnitude unit in size, there will be about ten times as many aftershocks.

Like everything in nature, none of these rules are hard-and-fast. Aftershocks have been less strong, as strong, or stronger than the original quake, dissipating more quickly or more slowly. For example, the New Madrid quakes of 1811–1812 never dissipated over a period of two and a half months (though some scientists believe these quakes followed Omori's law).

The biggest mystery facing scientists is how aftershocks are triggered, and why such shakings can occur a day or two—or longer—after a quake. Some seismologists suggest that these quakes are triggered by static stress resulting from the movement of the crust—in other words, stress on nearby faults as the crust responds to and rearranges from the initial quake. But not everyone agrees. Researchers of a recent study at the University of California, Santa Cruz believe that aftershocks are associated with the shaking from the actual main shock, not from added stress on nearby faults. They found that with increasing distance from the

> ## Creepy Movements
>
> There is one more type of movement along a fault that seems to be just as important as the sudden jerks we associate with earthquakes: *fault creep*. This movement is just what it sounds like—the slow, though sometimes rapid, slippage along active branches of a fault line. It is also called *seismic creep* to distinguish it from slumping of rock or soil along steep slopes; *aseismic creep* because the movement triggers only microearthquakes, if any movement at all; and *aseismic fault slip*. Whatever the term, creep occurs during the time between large stress-releasing quakes on a fault, or even as "afterslip" in the days to years following an earthquake.
>
> Why a fault slowly slips isn't totally understood. The best explanations seem to be low frictional strength on the fault, normal stresses on a fault line, and even elevated fluid pressures. In most cases, the creep movement seen at the surface usually depends on the rate of strain on the fault. If there is no creep, it is usually an indication that the fault is locked.
>
> Why the interest in creeps? Scientists realize that these movements are good indicators of the sheer strain on a fault. In many places, knowing how creep rates vary along faults has important implications in terms of forecasting the timing, location, and, some say, potential magnitude of future quakes. For example, scientists know that creep rates are sensitive to stress changes in the crust caused by moderate to large quakes on neighboring faults in the region. These stress changes can act either to advance or delay the timing of future quakes along the fault. Measuring creep rates over time can even be instrumental in estimating a fault's locking depth—which may then be used to interpret where the greatest slip will occur during a quake.

initial quake, the number of aftershocks dropped steeply—over a range of under a tenth of a mile to up to 30 miles (a quarter of a kilometer to 50 kilometers). This suggests that the same triggering process operates over the entire range—even with smaller initial quakes. And interestingly enough, the aftershocks fall off with distance in the same way that seismic waves do—thus, the researchers believe that the chance of having an aftershock depends directly on the amplitude of the shaking.

When do aftershocks "officially" end after a moderate to major quake? In general, an aftershock sequence ends when the quaking drops back to background levels, which in most cases means no quakes are felt at all—at least until another quake occurs. In active volcanic regions, the end to aftershocks doesn't mean the shaking ends, but that the number of quakes goes back to the usual constant background shaking.

CHAPTER 4

The Measuring Mess

> On August 21, 2003, a 7.2 earthquake occurred off New Zealand's southwest coast. It was felt in Sydney, but not in Wellington, New Zealand, 600 kilometers away. There is a layer of water in the ocean at 1,000 metres depth (called the SOFAR channel) in which sound waves generated by earthquakes can travel for thousands of kilometers. These sound waves traveled across the Tasman Sea in this layer. The effect in Sydney was like a magnitude 3.0 earthquake only 50 kilometers off the coast.
> —From the Commonwealth of Australia, Geoscience Australia

Instruments to measure modern earthquakes are much more sophisticated than the one made by Chinese philosopher, mathematician, astronomer, and geographer Zhang Heng around 132 AD, during the Later Han Dynasty. Though his original construct, called *houfeng didong yi*, or "instrument for inquiring into the wind and shaking of the earth," did not survive, the best reconstruction shows a dragon-head-festooned bronze jar with a central pendulum inside. There has been some scholarly disagreement as to the actual mechanics behind the instrument, but in the most accepted suggestion, when a quake struck, the pendulum would push a metal ball inside, dropping it out of one of the dragon's mouths and into an equally decorative frog's mouth below. The direction of the quake—which is all this ornamental apparatus would measure—was indicated by which of the dragon heads dropped the ball.

Obviously, theories of earthquakes have advanced since Heng's contraption was devised. Today, there are two ways of looking at all that shaking: by the intensity, or the observed effects of an earthquake on its surroundings, and by magnitude, or the measure of the seismic energy released by the earthquake.

MEASURING INTENSITY
First True Measurements

Intensity measurements have been the most useful over time, allowing scientists to take historical records and interpret the intensity of a quake based on what physically happened to an area. Such scales allow the interpretation of quakes in areas that have no seismograph or other instrumentation to measure local quakes. And even though intensity scales are as close as scientists can get in some regions, there can be limitations: sometimes there are no chimneys to crumble or windows to rattle.

The road to earthquake intensity scales was not paved with good intentions, but with constant scientific tweaking. The earliest scale was developed by P. N. G. Egen to describe the 1828 Belgian earthquake; much earlier, but not given as much credit, Italian architect Pompeo Schiantarelli developed a quake intensity scale, using simple symbols to represent damage after the Calabrian earthquake of 1783. But neither Egen's nor Schiantarelli's innovation caught on, and it took until the latter nineteenth century before the use of earthquake intensity to describe the shaking became widespread.

The idea of using physical damage caused by a quake to measure its intensity came from Michele Stefano Conte de Rossi of Italy and François-Alphonse Forel of Switzerland, two scientists who published their intensity scales in 1874 and 1881, respectively. They combined their efforts, in 1883 publishing the resulting Rossi-Forel scale, a 10-element scale and the first to be used internationally.

The 1883 version is as follows:

I. Microseismic shock—recorded by a single seismograph; the shock is felt by an experienced observer.
II. Extremely feeble shock—recorded by several seismographs of different kinds; felt by a small number of persons at rest.
III. Very feeble shock—felt by several persons at rest; strong enough for the direction or duration to be appreciable.
IV. Feeble shock—felt by persons in motion; disturbance of movable objects, doors, windows, cracking of ceilings.
V. Shock of moderate intensity—felt generally by everyone; disturbance of furniture, ringing of some bells.
VI. Fairly strong shock—general awakening of those asleep; general ringing of bells; oscillation of chandeliers; stopping of clocks; visible agitation of trees and shrubs; some startled persons leaving their dwellings.
VII. Strong shock—overthrow of movable objects; fall of plaster; ringing of church bells; general panic; no damage to buildings.

VIII. Very strong shock—fall of chimneys; cracks in the walls of buildings.
IX. Extremely strong shock—partial or total destruction of some buildings.
X. Shock of extreme intensity—great disaster; ruins; disturbance of the strata; fissures in the ground; rock falls from mountains.

Not long after, Italian scientist Giuseppe Mercalli came on the scene. At first, he modified Forel's initial scale, changing the number of intensities to 6. He improved the Rossi-Forel scale in 1902, keeping the 10 degrees but making the divisions more detailed. Certain scientists were still not satisfied, and by 1904, Italian physicist Adolfo Cancani decided to expand the scale to 12

Image from Paso Robles, California, on January 25, 2004. These damaged historic structures and a vehicle were seen in downtown Paso Robles after the 6.5 San Simeon earthquake. *Courtesy of FEMA.*

degrees. Gaps in the new scale's descriptions became its downfall, and by 1912, August Sieberg had developed a new intensity scale often referred to as Sieberg's Mercalli-Cancani scale, one that became the foundation for all the modern 12-degree scales. (He also developed a tsunami intensity scale in 1927 that was, no surprise, further tweaked by Nicholas Ambraseys in 1962.) Sieberg's intensity scale was revised two more times, and was eventually modified once again and published under the name Mercalli-Cancani-Sieberg scale, or MCS scale—the one that is still used in southern Europe today.

By 1931, the MCS scale was translated into English—American seismologists Harry Wood and Frank Neumann, the translators, completely ignoring the *C* and *S* of the scale, calling it the Modified Mercalli (MM) intensity scale. As usual, certain scientists could not leave well enough alone, and in 1956, American seismologist Charles Richter published yet another modification of the MM scale. Not wanting to have his name attached to the document—and keeping with what seemed to be the tradition of giving Mercalli all the credit—Richter merely called it the Modified Mercalli scale of 1956, or MM56.

A general synopsis—and one of many modifications—of the Modified Mercalli intensity scale follows, with some Richter scale equivalents (read on for more on Charles Richter and the Richter scale):

MM I—Not felt by humans, except in especially favorable circumstances, but birds and animals may be disturbed; reported mainly from the upper floors of buildings more than ten stories high.

MM II—Felt by a few persons at rest indoors.

MM III—Felt indoors, but not identified as an earthquake by everyone (equivalent to less than magnitude 4.2 on the Richter scale).

MM IV—Generally noticed indoors, but not outside.

MM V—Generally felt outside, and by almost everyone indoors; most sleepers awakened; a few people frightened (equivalent to less than magnitude 4.8 on the Richter scale).

MM VI—Felt by all; people and animals alarmed; many run outside; difficulty experienced in walking steadily (equivalent to less than magnitude 5.4 on the Richter scale).

MM VII—General alarm; difficulty experienced in standing; noticed by drivers of motor vehicles; trees and bushes strongly shaken (equivalent to less than magnitude 6.1 on the Richter scale).

MM VIII—Alarm may approach panic; steering of vehicles affected.

MM IX—General panic; some houses collapse and pipes break open (equivalent to less than magnitude 6.9 on the Richter scale).

MM X—Most masonry structures destroyed; ground cracks profusely and liquefaction and landslides widespread (equivalent to less than magnitude 7.3 on the Richter scale).

MM XI—Wooden frame structures destroyed; most buildings and bridges collapse, along with most other infrastructures (equivalent to less than magnitude 8.1 on the Richter scale).

MM XII—Damage virtually total; trees fall, ground rises

and falls in waves (equivalent to greater than magnitude 8.1 on the Richter scale).

Though there were several more attempts to modify the scale again in the 1970s, they never caught on. At last count, there were more than a dozen scales called Modified Mercalli scales—and most of them were not even close to resembling Mercalli's original. In the United States, the 1931 version of the Modified Mercalli scale—published by American seismologists Harry Wood and Frank Neumann, mentioned above—is most often used. Like most of the other MM scales, it measures the various levels of intensity using a Roman numeral listing—from the Roman numeral I for the low end of the quake scale to XII for the most catastrophic quake imaginable.

This image of San Francisco and the San Mateo County coast, showing large slides north of Fort Funston, is from the Loma Prieta, California, earthquake on October 17, 1989. *Courtesy of USGS.*

The Omori Seismic Intensity Scale

In the late 1800s, it seemed as if every earthquake scientist decided to develop his own earthquake intensity scale. Though the more famous one is the Mercalli, there were others in between, including Fusakichi Omori's own scale, based on the typical Japanese structures of his time (it's interesting to note that because of the way the Japanese built their homes and businesses, Omori's level I is equal to a Mercalli intensity of VI):

I. The shock is rather strong, so much so that it generally induces people to escape from their houses into the open. The walls of badly constructed brick houses crack slightly and some parquet falls down; ordinary wooden houses are shaken in such a degree that they loudly creak; furniture is overturned; trees are visibly shaken; the water in ponds and pools gets turbid, owing to the disturbance of the mud; pendulum clocks stop; some very badly built factory chimneys are damaged.

II. The walls in the wooden houses of Japan crack; old wooden houses get slightly out of plumb; Japanese tombstones and the badly constructed stone lanterns are overturned; in a few cases the flow of the thermal and mineral springs is changed; ordinary factory chimneys are not damaged.

III. About one fourth of the factory chimneys are damaged; badly constructed brick houses are partially or totally destroyed; some old wooden houses are destroyed; wooden bridges are slightly damaged; some tombstones and stone lanterns are overturned; Japanese sliding doors (covered with paper) are broken; the tiles of wooden houses are displaced; some fragments of rocks are detached from the sides of mountains.

IV. All factory chimneys are ruined; the majority of the ordinary brick houses are partially or totally destroyed; some wooden houses are totally destroyed; the wooden sliding doors are mostly thrust out of their channels; crevices from 2 to 3 inches (5 to 7.5 centimeters) wide appear in low and soft grounds; here and there the embankments are slightly damaged; wooden bridges are partially destroyed; ordinarily constructed stone lanterns are overturned.

V. All ordinary brick houses are very seriously damaged; about 3 percent of the wooden houses are totally destroyed; some Buddhist temples are ruined; the embankments are badly damaged; the railways are slightly contorted; ordinary tombstones are overturned; brick walls are damaged; here and there, large fissures from 1 to 2 feet (30 to 60 centimeters) wide appear along the banks of the watercourses. The water of rivers and ditches is thrown on the banks; the contents of the wells are disturbed; landslides occur.

VI. The greater part of the Buddhist temples are ruined; from 50 to 80 percent of the wooden houses are totally destroyed; the embankments are almost destroyed; the roads through paddy fields are ruined and interrupted by fissures in such a degree that traffic by animals or vehicles is impeded; the railways are very much contorted; great iron bridges are destroyed; wooden bridges are partially or totally damaged; tombstones of solid construction are overturned; fissures some feet wide appear in the soil and are sometimes accompanied by jets of water and sand; iron or terra cotta tanks embedded in the ground are mostly destroyed; all low-lying grounds are completely convulsed horizontally as well as vertically in such a degree that sometimes the trees and all the vegetation on them die off; numerous landslides take place.

VII. All buildings are completely destroyed, except a few wooden constructions; some doors or wooden houses are thrown over distances from 1 to 3 feet (a third of a meter to a meter); enormous landslides with faults and shears of the ground occur.

Scales from Afar

Not everyone around the world uses the same intensity scales. Certain countries insist on their own methods, with one version or another usually used in remote areas that have no devices to measure quakes. Overall, there are slight differences between the scales. For example, the Medvedev-Sponheuer-Karnik (MSK-64) scale is used by India, Israel, Russia, and the Commonwealth of Independent States. It was first developed by Sergei Medvedev (USSR), Wilhelm Sponheuer (East Germany), and Vít Kárník (Czechoslovakia) in 1964, based on the MCS and MM56 scales, as well as a previous scale developed by Medvedev. It uses Arabic numerals, starting from the number 1, or not perceptible (not felt, registered only by seismographs; no effect on objects; no damage to buildings), to 12, or very catastrophic (all surface and underground structures completely destroyed; landscape generally changed; rivers change paths; tsunamis).

The shindo scale used in Japan and Taiwan describes the degree of shaking at a point on the Earth's surface (also called the ground acceleration). The scale runs from 0 (insensible, or imperceptible to people) to 7 (ruinous earthquake, with wall tiles, most buildings, windowpanes, and concrete-block walls collapsing). Thus, Japanese quakes are often reported as, "shindo 4 in Tokyo, shindo 3 in Hiroshima," and so on.

The scale used throughout Europe, the 12-degree European Macroseismic Scale (EMS), has quite a history, too. It started in 1988, when the European Seismological Commission (ESC) decided to begin a radical revision of the MSK scale, starting a working group called, logically, "Macroseismic Scales." No one ever said science was fast, and it took years to change the scale: by 1992, the revised version was finalized, and it was ratified at the ESC meeting in 1996, with the final version published in 1998. The differences were immediately obvious. This was the first intensity scale that actually came with "instructions" for use—not the usual obtuse definitions from the other scales that were often misinterpreted by the observer (see the above MM intensity scale). The resulting charts even contain illustrations and pictures depicting the various levels of damage—a way to standardize people's interpretations of quake damage.

Recent Rockings

What are some familiar recent earthquakes? It's like a litany of death and destruction:

- the Pakistan earthquake of October 8, 2005, a magnitude 7.6 quake that killed an estimated 86,000 people;
- one of the highest magnitudes ever recorded—a 9.0 that occurred on December 26, 2004, off the west coast of Sumatra, and was responsible for a giant tsunami that killed an estimated 230,000 people in the region;
- the March 28, 2005, magnitude 8.7 earthquake in northern Sumatra that killed an estimated 2,000 people;
- a quake on December 26, 2003, in Bam in southeastern Iran, magnitude 6.5, with more than 31,000 killed;
- the January 26, 2001, India, magnitude 7.9 quake in which at least 2,500 were killed, though estimates put the death toll as high as 13,000;
- a quake on August 17, 1999, in western Turkey, magnitude 7.4, in which 17,000 were killed;
- a quake on July 28, 1976, in Tangshan, China, magnitude 8.2 to 7.8 (no one agrees), in which at least 250,000 people died, with some estimates higher. This quake was one of the deadliest in recent times.

The list will never be complete. Every year, on average, there is at least one major quake somewhere around the world. You can start adding your own list for comparison—an exercise that will help you realize the constant threat we face from such quakes.

MEASURING MAGNITUDE
Magnitude Rules

In between all the modifications of the Modified Mercalli intensity scale, some scientists believed there had to be a better way to describe a quake than by its physical effects in a particular place. This alternative (or addition) to the intensity scales eventually became known as magnitude—a more standard way of measuring a quake based on the amount of energy it released.

The history of measuring earthquake magnitude is just as convoluted as sorting out quake intensity scales. One of the first earthquake magnitude scales is also the most familiar: the Richter scale, developed, as described above, by American seismologist Charles F. Richter in 1935 in collaboration with German scientist Beno Gutenberg, at the California Institute of Technology. Also called the Richter magnitude test scale—or local magnitude scale, ML—it assigned a single number to quantify the size (energy) of an earthquake.

The listing is a base-10 logarithmic scale, or the amplitude of the seismic waves increased by a power of ten as the numbers increase; the same unit increase in magnitude corresponds to an increase of 32 times the quake energy. Inspired by the magnitude scale used by astronomers to measure the brilliance of stars and other space objects, Richter chose the word *magnitude* and numbers that ranged from 0 to 9, with no true upper limit. (It was hoped there would be no magnitude 10 quake, ever, as such a quake's destruction would reach all the way around the world.) The scale itself was designed only for a local area of Southern California and was to be used in association with a particular instrument—the Wood-Anderson torsion seismometer. In order to prevent negative magnitudes, Richter chose 0 as the beginning point, which represented an earthquake that would have a maximum combined displacement of a fraction of an inch (1 micrometer) on this special seismometer sitting around 62 miles (100 kilometers) from the quake's epicenter.

Over time, scientists realized that there were several problems with the Richter scale, especially since it becomes extremely imprecise when measuring quakes in the magnitude 8 and 9 range. In particular, in this range, the numbers break down, often yielding the same estimates for events that clearly differ in magnitude.

Several possibilities soon emerged, and a decade ago, seismologists began to devise other ways of looking at quakes. It may seem like overkill, but in reality, each method works only for a limited range of magnitudes and with different types of seismometers. They include the body-wave magnitude (Mb, ranging in magnitude from 4 to 7, based on the amplitude of P body waves, and usually most appropriate for deep-focus quakes); coda duration magnitude (Md, based on the amount of time the shaking occurs as measured by the decay in wave amplitude, usually measuring below a magnitude 4); and the surface-wave magnitude (Ms, used for distant earthquakes based on the amplitude of Rayleigh surface waves). The results of each measurement are similar to the Richter scale, but because each value is based on the measurement of one part of the seismogram, they do not take into account the overall power of the earthquake's source; and, like the Richter scale, they often undervalue magnitudes of large events. But overall, they add more information to complex earthquake equations.

To take up the slack, especially in the higher-magnitude earthquakes, scientists developed an even more useful and accurate scale (even though the Richter scale is still champion in the media). It is called the Moment Magnitude (sometimes seen incorrectly as Magnitude Moment; written as M_w) scale, developed by seismologists (again at the California Institute of Technology) Tom Hanks and Hiroo Kanamori. This scale, with its roots in the physics concept of "moment," is based on several parameters of the quake, including the size of the fault on which the earthquake occurs, energy from the quake, and the amount the Earth's crust slips. In particular, it provides clues to the physical size of quakes, as well as the amount of energy released.

The moment magnitude can be derived not only from a seismogram, but by taking other geodetic measurements. It's calculated in part by multiplying the area of the fault's rupture surface by the distance the earth moves along the fault. The moment results are then converted to a number similar to other earthquake magnitudes by a standard formula. It provides an estimate of earthquake size that is valid over the complete range of magnitudes—something that was always missing in other magnitude scales.

Overall, the Richter scale and the Moment Magnitude do not differ significantly—but it depends on the quake. For example, the 1906 San Francisco quake measured 8.3 on the Richter scale, M_W 7.7 on the Moment Magnitude scale. The 1964 Prince William Sound, Alaska, quake was 8.4 on the Richter scale, M_W 9.2 on the Moment Magnitude scale. And the 1994 Northridge quake measured 6.4 on the Richter scale, M_W 6.7 on the Moment Magnitude scale.

Alas, the use of a defining, standard magnitude has a long way to go, especially in the media, which is why we often read or hear differing magnitudes (and scales) for various earthquakes. Many historical records and earthquake reports use the Richter scale, and the transition to Moment Magnitude will take a while. Most scientists hope that the measuring mess for earthquake intensity, magnitude, and any other function will one day be standardized.

The Scale of It All

Though it seems as if there is an abundance of magnitude scales, there is still somewhat of an agreement as to shaking—at least between some scales. The following listing is used by the United States Geological Survey (USGS), and shows the approximate Moment Magnitudes for the major scales, with the possible effects (though it should be noted that this is a general list):

Great is equal to a magnitude greater than 8.0;
Major is equal to a magnitude between 7.0 and 7.9;
Strong is equal to a magnitude between 6.0 and 6.9;
Moderate is equal to a magnitude between 5.0 and 5.9;
Light is equal to a magnitude between 4.0 and 4.9; and
Minor is equal to a magnitude between 3.0 and 3.9.

Anything less is considered to be from very minor to a microquake. In fact, the USGS does not use the Moment Magnitude scale for quakes less than magnitude M_W 3.5.

Measuring Machines

How do scientists gather all these magnitude numbers to crunch? We've come a long way since Heng's dragon-mouth, ball-dropping container. Today, networks of seismic stations measure ground movement at a certain site, and if the quake is strong enough, stations can capture the evidence of a quake from far away. Data are recorded on seismographs; at the heart of the instrument is a seismometer, in which a pendulum or mass is mounted on a spring that moves during a quake. When the ground shakes, the seismograph produces a seismogram, the classic squiggles on paper surrounding a slowly moving drum that we've all seen on television or in movies; more modern seismograms record data digitally, sending necessary data to interpret an earthquake's focus, waves, and epicenter directly to a computer.

Whether a paper-covered or virtual "drum," the incoming data reveal a horizontal axis of time measured in seconds; the vertical axis shows ground displacement, usually measured in millimeters. (The seemingly persistent, occasional wavy lines when there are seemingly no quakes are classified as "noise," with some instruments sensitive enough to measure a passing person or garbage truck rumbling by.) Using geometric gymnastics, the data from each seismic station close to and far from the quake determine the source and strength (magnitude) of the quake.

The best part about these machines is their transportability. Currently, and no doubt in the future, more seismometers will be placed in various spots around the world, most of them located

---

History of the Shakes

One of the more fascinating aspects of earthquakes is the historical study of the phenomenon. What did scientists of long ago know about earthquakes? Here are a few former theories about why we have earthquakes–contrasted with what we know today:

Then: Greek philosopher Aristotle (384-322 BC) developed a fanciful explanation for quakes, albeit rational for his time. He believed that strong winds blew through caves inside the Earth creating "effects similar to those of the wind in our bodies whose force when it is pent up inside us can cause tremors and throbbings."

Now: We know that wind in caves can't possibly have an effect on the movement of the crust. And though Aristotle got the "inside the Earth" part right, he probably would never have fathomed that there was moving liquid rock tens of miles below his feet.

Then: In 1793, Benjamin Franklin proposed a solution for geologist John Michell's theory that quakes were caused by shifting masses of rock many kilometers below the surface. Franklin imagined that the

on land, some in the oceans. These networks of instruments will watch major and minor quakes, allowing scientists to see data where only Mercalli measurements were available before. This will fill in the gaps not only about plate boundary earthquakes, but also about quakes in the wrong places.

How many quakes do earthquake instruments and networks measure annually? At one time, scientists believed there were hundreds of thousands of earthquakes around the world every year. Thanks to more seismograph stations and better technology to collect data about the most minute quakes, we now know there are actually *millions* of earthquakes per year. These seisms range on average from 19 major to more than 2 million minor ones every year. On average, that's over 6,000 per day, 18,000 per month, and 250 per hour—all statistics that represent ballpark numbers only.

> "internal part [of the Earth] might be a fluid more dense, and of greater specific gravity than any of the solids we are acquainted with; which therefore might swim in or upon that fluid. Thus the surface of the globe would be a shell, capable of being broken and disordered by any violent movements of the fluid on which it rested."
>
> *Now:* Franklin hit the nail right on the head—he was one of the first to suggest that the shifting rock (his "shell") would move and create quakes. It would take more than a century to revive the idea and take it to its fruition—today's plate tectonics.
>
> *Then:* Scientists in the latter part of the nineteenth century proposed that after the Earth formed, it cooled—and the planet shrank, much like an apple drying in the sunshine. This shrinking caused the cracks in the planet, all of which would move and quake as the Earth shrank more and more.
>
> *Now:* We know that the Earth is cooling down, but few scientists believe we are shrinking; in fact, they know that the size of the Earth has not changed significantly in the past 600 million years—and probably not even since its formation 4.56 billion years ago.

CHAPTER 5

Quirks of Quakes

The phenomenon of ball lightning, called "an unsolved problem in atmospheric physics" [by Stenhoff in 1999], may be another example of an unusual phenomenon. Ball lightnings are free-floating volumes of ionized air that detach themselves from the ground. According to eyewitness reports, small ball lightnings have entered into rooms through windows, often without leaving a trace or any cracks in the glass, or have entered rooms through telephone jacks and electric sockets. While drifting through the air, the balls reportedly produce a faint hissing sound. They explode with a bang after a few seconds and leave behind a smell of ozone. Such balls of ionized air seem to appear before or during large thunderstorms and before or during seismic activity. In the case of earthquakes, these plasma balls may detach themselves from the ground when clouds of p-hole charge carriers arrive at the Earth's surface, leading to high electric fields, similar to the fields measured during periods of intense thunderstorm and lightning activity.
—From *"Rocks That Crackle and Sparkle and Glow: Strange Pre-Earthquake Phenomena" by Friedemann T. Freund, NASA Ames Research Center, in* Journal of Scientific Exploration, *2003*

Not every earth movement is the result of an earthquake—and sometimes certain physical events can actually *cause* a quake, but not the type we hear about on the evening news. Then there are the strange happenings or quirks before, during, and after certain quakes, ranging from frost quakes and light shows to noises that sound like the cavalry coming over the hill. Here are a few freaky stories about some rare quakes and their effects—some near known quake-prone zones, some in the wrong places.

QUAKE QUIRKS CLOSE-UP
Booming Quakes

Sound travels in pretty much any medium, except the vacuum of space. It travels through water, thick and thin air, and even the Earth's crust. That's why so many people, especially in the Northeast and along the East Coast of the United States, claim to hear occasional "booms" during an earthquake. And though many of them are associated with human noise—from explosions, large vehicles passing, or even a sonic boom—scientists believe some of the racket may actually be shallow earthquakes too small to be recorded but large enough to be heard and felt by people nearby.

For example, the New Madrid quakes from 1811 to 1812 were also famous for their artillery-like booms (see more about the New Madrid quakes in chapter 6). Another such quake occurred in 2001 in a Spokane, Washington, earthquake swarm. (Earthquake swarms are a series of minor quakes—none of which are identified as the main shock—occurring over a certain area and time; for more about swarms, see chapter 9.) The small quakes unnerved the residents, but it was even more eerie because the tremors were also accompanied by "booming sounds." Scientists believe these quakes were shallow, only about one to two miles (a half to one kilometer) deep. The booming was probably caused by higher-frequency vibrations just below the surface. In deep quakes, the vibrations dissipate before they reach the surface; in shallower quakes, the vibrations are felt and heard—or not felt, but only heard.

This may explain why some people claim, "I heard it coming," or, "I heard the quake before it hit the area." Scientifically speaking, there is little chance that someone can hear the quake before it hits. Sound travels in the atmosphere about 1,100 feet (335 meters) per second; when seismic waves leave the source of an earthquake, they are moving nearly 20 times the speed of sound, which means the quake waves hit before any sound reaches the observer.

So what causes the booming noises? In one experiment, researchers recorded magnitude 2.0 and 3.0 earthquakes, measuring the arrival of the P-wave on a seismograph. They noted that a sound arrived before the S-waves were recorded. The scientists concluded that the booming sounds were actually the arrival of the P-wave. As the P-waves expand out from the focus of the quake, the particle motion in the wave is back and forth along the direction of wave propagation, like hitting a long steel rod with a hammer on one end and picking up the vibration at the other end. This is similar to the particle motion of a sound wave traveling through the atmosphere—meaning the P-wave essentially is a sound wave traveling through the Earth. When the P-waves emerge at the surface, a fraction of them are transmitted into the atmosphere as sound waves. That is why, when these waves do hit, many people close to the epicenter often hear the high frequency of the seismic waves—a noise often described as thunder, a distant locomotive, or cannon fire. Others have described the noise as an explosion followed by a train rushing past.

Bright Lights, Earthquake Lights

Other strange occurrences around many quakes are earthquake lights, reportedly resembling the quick-moving, white to blue, ghostly features of an aurora; they can also appear as a flake or ball, or even electric sparkles. They were long thought to be a myth perpetrated by superstition, with reports occurring before, during, or after a quake. One of the first records of such lights was written around 373 BCE in ancient Greece: the huge "columns of flame" appeared just before a massive earthquake destroyed the cities of Helike and Boura.

For many centuries afterward, claims of earthquake lights were still ignored—until evidence emerged in the form of photographs taken during the Japanese Matsushiro quake swarm (reportedly more than 60,000 quakes with magnitudes up to 5.5) between 1965 and 1967. There were also the electromagnetic waves that occurred during the great Chilean earthquake of 1960—waves that even interfered with radio transmissions during the quake. Over 100 sightings of earthquake lights were reported in Quebec, Canada, between November 1988 and January 1989, during the Saguenay earthquake sequence. At one point, a rapidly propagating sheet of light—accompanied by a bristling noise in the trees—was witnessed 12 miles (19 kilometers) from the epicenter just before the largest shock. Even more telling were the more than 20 sightings of various lights during the 1995 Kobe, Japan, earthquake—streaks of white and blue to orange rising into the sky.

Since that time, many explanations for the lights have been suggested, but no one knows the real cause. Here are just a few educated guesses: a piezoelectric effect—scientists know that a mineral such as quartz can generate voltage when mechanical stress is applied, and if the quake occurs around a quartz-bearing rock, it may create sparks, much like striking flint to start a fire; frictional heating, in which any movement of rock will create heat—and if the rock in a quake moves a great deal, it may create light; sonoluminescence, in which a strong enough sound wave—possibly like the one that accompanies a major quake—can cause gas in a liquid to collapse, and since most rock contains gases along with liquids (like water), it is possible that short bursts of light from the imploding bubbles could result; and triboluminescence, or another "luminescence" phenomena, in which light is generated when certain rock is crushed, scratched, or rubbed (in fact, diamonds often exhibit this glowing effect when they are being faceted or cut).

Quaking Months?

Want to know which month wins in the earthquake department? According to the USGS, March seems to be the time for the "great" category of quakes (magnitude 9.0 and above) in the United States. For example, based on USGS seismic data, March saw the two largest earthquakes ever recorded in the country's history: on March 28, 1964, Prince William Sound, Alaska, experienced a Moment Magnitude M_W 9.2 event that killed 125 people and caused $311 million in property damage. On March 9, 1957, the Andreanof Islands, Alaska, felt a Moment Magnitude M_W 9.1 temblor that destroyed two bridges on Adak Island, damaged houses, and left a 15-foot (4.5-meter) crack in a road. But its real damage came in the form of a tsunami: a 50-foot (15-meter) wave hit Scotch Cap; the tsunami continued to Hawaii, destroying two villages and causing $5 million in property damage on Oahu and Kauai islands.

Weathering Quakes?

Is there a connection between quakes and the weather? If there is, the chances of a quake occurring out of nowhere increase dramatically. There is even a saying: "It's hot and dry—that means earthquake weather," especially in quake-prone areas such as California. One reason for the propensity to link the two is that people tend to notice a quake if it seems to fit a pattern—not if it doesn't fit any pattern at all. And since it's usually hot and dry in California, there's a good chance for a quake during such weather.

Alas, no matter how you look at it, there is really no direct connection between a quake and rain, snow, or heat. Earthquakes occur miles below the surface—too far for quakes to form from even a major storm like a hurricane or tornado.

Not that weather has been ignored when it comes to quakes. For example, India is no stranger to earthquakes; this chunk of land smashed into the Asian continent, creating the Himalayas. So when one hears about one of the most famous earthquakes being the October 11, 1737, Calcutta quake, a shaking responsible for the loss of more than 300,000 lives, making it rank with the most destructive in recorded history—most people take notice. But in 1984, Roger Bilham, then at the University of Colorado, Boulder, decided to look into the records. Despite the mention of the humongous quake in many old geology books—not to mention the popular press—it appears that the quake was not a quake at all. The reality may be that the ferocious devastation was caused in fact by a massive cyclone.

Part of the evidence comes from merchants returning to Europe from Bengal around six months later. There were reports of a destructive cyclone at the mouth of the Ganges River in the

Bay of Bengal, one chronicle mentioning "15 inches of water in 6 hours" while winds, thunder, and torrents of rain damaged homes and other structures. Trees were blown down and ships in the harbor were carried out to sea. The official records from one merchant, the East India Company, mentioned the disaster and around 3,000 fatalities—but not a word about an earthquake. There appear to be no official reports from the Indian government about an earthquake that occurred in the region during this time. Add to this the recorded population of Calcutta at that time at less than 20,000—and the mention of "drowned crocodiles" (alligators do not populate this part of India)—and the report of a quake becomes suspect.

Where did this fictitious information—probably one of many such instances over the centuries—originate? It may have been the media: an account in a 1738 London magazine has been used over and over as a primary source. Maybe it was rumors; maybe the reports of such massive destruction caused the writer of the article to assume the worst. Of course, there is a chance that the two natural disasters struck at once, with the hurricane the more ferocious of the two. But most historians believe the damage was actually caused by the hurricane winds and resulting flooding. And if the ground did shake, it was probably minor compared to the destruction of the cyclone.

Quakes in the Cold
In the coldest regions of the world, many areas are covered by permafrost, or semipermanent frozen ground. When the "active" surface layer thaws in the summer, the mix of meltwater, soil, and rock can flow, causing the surrounding area to

> ### Changing Ancient Cultures with Quakes
>
> One of the more overlooked fields of geology is the study of ancient seismites—the evidence in sedimentary rock that points an accusing finger at an earthquake. Of course, the study of ancient quakes is not found only in the rock record. It is also gleaned from our own past. For example, the great Lisbon earthquake of 1755 and its associated tsunami—which struck while the churches were celebrating All Saints' Day—had a dramatic effect on western European culture, shocking the likes of Voltaire. Furthermore, it caused an intellectual revolt against the Church—many people thought the quake and tsunami were manifestations of a truly angry God punishing the Church—helping to start the Age of Reason. Historians also point to certain civilizations that came and went in Asia Minor, shaken not only physically but culturally after catastrophic earthquakes. And, of course, not all this death, destruction, and cultural change occurred around tectonic plate boundaries—some catastrophes occurred in places where plate borders lay thousands of miles away.

become waterlogged and sink. Add to this the fact that global warming may affect the extent and solidness of permafrost areas, and it's easy to see why movement from thawing permafrost can become a major concern.

In general, frost action, a type of physical weathering, can move the ground, but usually not enough to produce any quaking. Water commonly collects in cracks and wedges everywhere on Earth. In a cold region—especially during the winter and early spring—water seeping within cracks can pry apart rocks as it freezes (water expands by 9 percent as it turns to ice) and thaws. The eventual crumbling of the rock can take a few to hundreds of years, depending on the rock type and the amount of precipitation in a region. Frost heave is caused by a similar process: water in the soil creates ice lenses that expand, causing the ground to rise. The amount of heaving depends on how much water is available to freeze; if more water is added, the lens can continue to grow. Such heaving affects roads, sidewalks, and foundations—or even a garden, heaving up rocks buried in the soil.

It's true that frost quakes and frost heave have never caused cities to fall or civilizations to crumble. The biggest danger is the falling of rock from cliffs and steep slopes that have been weathered by ice. But people have reported another possible type of quake from frozen ground—a phenomenon called a frost quake, or cryoseism that has been reported in upstate New York, Maine, Vermont, Massachusetts, and Connecticut. These cryoseisms are most often caused by the sudden deep freezing of the ground. They typically occur during the first cold snap of the year, when temperatures drop from above to below freezing—and particularly if there is no insulating snow on the ground.

The resulting localized shaking does not travel far, as cryoseisms do not carry as much energy as true earthquakes. But because they occur at the ground surface, they have been known to shake people from a sound sleep or crack a foundation. For example, in Maine, one such frost quake occurred on February 25, 2003: a loud bang was heard. Residents found a crack 30 feet (10 meters) long in the snow crust, along with a crack in a concrete floor that ran the length of a house. Though there was no snow cover at the time, a drop in temperature that night to below freezing was typical for such a shaking. Another occurred on January 28, 2005, in which a "good jolt" was felt and small objects vibrated; still another occurred on January 30 the same year—there was a loud noise, the ground shook—and a resident even went outside to see if his car had exploded.

> ## Can Skyscrapers Cause Quakes?
>
> The Taipei 101–in the Xin-Yi district of Taipei in Taiwan–is not only (at this writing) the tallest building in the world, but it is the center of a controversy. According to a theory presented by Cheng Horng Lin, a geologist from Taiwan, the Taipei Basin in which the Taipei 101 sits was once a stable region. Since the construction of the 1,667-foot- (508-meter-) high, 700,000-tonne building, there has allegedly been an increase in the number of quakes there. Lin claims the reason is the huge edifice–the building's structure and weight putting so much stress on the ground below that an earthquake fault (or faults) may have become active again.
>
> But not everyone agrees–especially those who want to build even higher skyscrapers in the future. Many scientists point out that Taipei is already in a known active tectonic region–thus the number of quakes changes and shifts over time. And the biggest problem with Lin's theory is that the pressure caused by the tallest building is not enough to affect the rock layers at the depth at which most quakes originate. This building stresses rock only to about 6 miles (10 kilometers) underground, not the average 62 miles (100 kilometers) for major quakes.

QUAKE QUIRKS FROM A DISTANCE
Lunar Quakes

The Earth is not the only planetary body in the solar system to experience earthquakes. Thanks to data from seismometers placed by the *Apollo* 12, 14, 15, and 16 astronauts at their respective landing sites, we now know the Moon has (obviously) moonquakes. More than 12,000 quakes were recorded in the eight years of data gathering. And even though the seismic activity is only about a hundred-millionth that of Earth's, if we go back to colonize our closest space neighbor, astronauts will definitely have to pay attention to the potentially shaky ground.

The Moon shakes may not occur for the same reasons as the Earth's quakes—the Moon does not have plate tectonics—but the shaking does occur. Many of the quakes are deep, about 435 to 746 miles (700 to 1,200 kilometers) below the surface, and are probably caused by tidal forces between the Earth, Moon, and Sun—a sort of ground tide. Other quakes are the minor vibrations from meteorite and artificial satellite strikes on the surface; still others, called thermal quakes, are probably the crust's response as the surface expands and contracts to the rhythm of extreme heating by the Sun—the sunrise and dawn. Finally, there are the shallow quakes, shaking only about 12 to 19 miles (20 to 30 kilometers) below the surface. No one really knows what causes these quakes, but most scientists point to landslides or the slumping of wall material in young craters.

But it's the shallow quakes that concern scientists the most. In fact, some have registered above magnitude 5.5 on the Richter scale, enough to move heavy furniture, crack plaster, and sway buildings; and because of the composition of the Moon's surface, these larger quakes can last for more than 10 minutes. The dry, cold, mostly rigid Moon acts like a huge metal tuning fork—once the vibrations get started, it's hard to stop them. Scientists now believe that if we do go back to the Moon, engineers creating the distant infrastructures will no doubt be taking tips from California's earthquake building codes.

E. T. Quakes

Is ours the only planetary body in the solar system that insists on having plate tectonics—and thus volcanoes and earthquakes?

According to data sent back by the various spacecraft visiting the inner solar system, it may be true. Mercury has long since cooled down, showing not even a trace of possible crustal plates on its surface. Mars, too, has had a lifetime of change, including cessation of volcanism and loss of any water. Numerous spacecraft have landed on the planet, but in all the reams of data, there has been no evidence of a Martian quake.

Venus has a better chance of having volcanoes and quakes, but not from the movement of plates—just from the planet not quite having cooled down yet. As on Earth, earthquakes may be a result of the volcanism, but no one has ever been able to land a craft long enough on the surface to confirm such shaking. After all, between the enormous pressure (10 times that of Earth) and heat (around 883 degrees Fahrenheit [477 degrees Celsius] at the surface), a craft is literally hard-pressed to gather any information.

The outer solar system may have more of a chance for quakes and eruptions. A major contender is Io, one of the larger moons of Jupiter. Images taken by many craft visiting the Jupiter system have revealed active volcanoes on that moon, probably from the gravitational (tidal) pull of the mother planet; no doubt the satellite has quakes, too. Another huge moon of Jupiter, Ganymede, is more likely to have convection and crustal plates, as spacecraft data have suggested that the moon has a liquid interior.

Alas, many of these planetary bodies will one day lose their ability to produce volcanic eruptions or shake the ground. Eventually the Earth, too, will lose so much of its interior heat that all convection in the mantle will cease. When that happens—billions of years from now—if anyone is left to notice, there will be no more volcanoes or earthquakes.

Quakes in the Wrong Places

CHAPTER 6

The New Madrid Disaster

What are we gonna do? You cannot fight it cause you do not know how. It is not something that you can see. In a storm you can see the sky and it shows dark clouds and you know that you might get strong winds but this you can not see anything but a house that just lays in pile on the ground—not scattered around and trees that just falls over with the roots still on it. The earth quake or what ever it is come again today. It was as bad or worse than the one in December. We lost our Amandy Jane in this one—a log fell on her. We will bury her upon the hill under a clump of trees where Besys Ma and Pa is buried. A lot of people thinks that the devil has come here. Some thinks that this is the beginning of the world coming to an end.

—*A January 23, 1812, eyewitness account of the New Madrid quake by George Heinrich Crist, who was at that time residing in north-central Kentucky, near the present location of Louisville. The language and spelling are authentically his.*

Stretching just west of the Mississippi River are the lands around New Madrid, Missouri, a region occasionally bothered by river flooding, mosquitoes as big as buses, towering thunderstorms, and even wandering tornadoes. The potential for another hazard sits beneath the population's collective feet: earthquakes—even though the small town lies thousands of miles from an active plate boundary. In fact, this area has produced enough ground shakes to be classified as a "classic case" of earthquakes in the wrong place.

Quakes occur along a long underground fault line called the New Madrid fault zone, which cuts through the region for more than 150 miles (241 kilometers). Scientists were made aware of this rip in the crust thanks to a series of major earthquakes that rocked the area between late 1811 and into early 1812—so far, the worst quakes in North America *and* the worst documented in a stable continental region. At least two of the massive earthquakes during those long winter months were more powerful than any recorded in the continental West—including anything the San Andreas Fault ever produced.

THE OLD NEW MADRID QUAKES
New Madrid Then
The old riverboat area of New Madrid lies almost in spitting distance from where the Ohio meets the Mississippi River. The town was originally within French-owned territory, founded as an Indian trading post in 1783 by French Canadian fur-trapping brothers Francois and Joseph LaSieur. They named it L'Anse à la Graisse, or "The Cove of the Grease" (for reasons of their own, no doubt). It passed under Spanish rule around 1763; by 1789, Revolutionary War veteran George Morgan was given control after entering into a colonization scheme with the Spanish ambassador to the United States. He renamed "the Grease" New Madrid after the famous city in Spain, with hopes of turning it into the capital of New Spain. As with all of the best-laid plans, the United States made a

big-ticket purchase in 1803—the Louisiana Purchase—that included the area of New Madrid. By 1821, it would reside in Missouri, the eleventh state to join under the federal Constitution.

At the start of 1811, the United States was a mere 35 years old, with a population of around seven million. Most people settled in cities hugging the East Coast, the number of individuals gradually thinning out toward the Mississippi River and beyond. James Madison and Vice President George Clinton were in power at a time of turmoil: certain elected officials—"war hawks" such as Henry Clay and Felix Grundy—were advocating war with both France and Great Britain (the United States would declare war on England late the next year); the Indian crisis was spreading, as certain tribal leaders began to resist the United States policy of obtaining land by "treaty"; and overseas, Napoleon Bonaparte was creating his own set of warfare and conquering problems.

Not to be outdone, nature started its own set of trials and tribulations: the New Madrid earthquakes of 1811 to 1812. For three months, the people around the region of New Madrid suffered a round of earthquakes that has not been experienced since. The three main shocks were immense, measuring over magnitude 7.0—and lasting from a minute to five minutes. The number of quakes measuring below magnitude 7.0 during those fateful months reached more than 2,000.

The initial quake shook the ground on December 16, 1811, at around two o'clock in the morning. The shallow quake originated a few miles below the surface west of present-day Blytheville, Arkansas, and south of New Madrid. According to accounts, the nearby towns of New Madrid (800 residents) and Little Prairie (200 residents; present-day Little Prairie Township below New Madrid) suffered almost total destruction. Standing was impossible during the quake, as the seismic waves caused the ground to visibly ripple. Trees were split, snapped off at the top, or toppled, roots and all; geysers of sand, water, and debris flew high into the air; and the land ripped open in spots, with some fissures reportedly more than 10 feet wide.

As residents began to believe the worst was over, another quake, nearly as strong as the first, hit under the present-day cities of Cooter and Steele, Missouri, between 7:00 and 8:00 A.M. This time, the few structures left standing in New Madrid and Little Prairie fell to the violence of the ground's motion. Surprisingly, there were few fatalities—probably only about a dozen—in either town. Modern dwellers of the New Madrid region should pay heed: scientists believe the low number of deaths may have been due to the sturdiness of the ubiquitous log cabins. The initial shock, strong as it was, was not enough to collapse the cabins, allowing for a quick escape. The latter shock probably tumbled the already unstable structures, but by then, most of the residents had evacuated.

Still another quake struck at around 11:00 A.M. beneath present-day Caruthersville (in today's Little Prairie Township), with less energy than the preceding shakes. But the lands around New Madrid were beginning to change. By the end of the third quake, Big Lake, just west of Blytheville, had formed—a new lake in one day.

The nearby Mississippi River didn't fare much better. Huge chunks of the Mississippi riverbanks toppled, pushing the water like an uncontainable tidal bore. These small-scale tsunamis quickly spilled onto lands downstream, inundating and eroding stretches of the surrounding banks. Islands rose in the river, while others disappeared, and waterfalls formed from the sudden uplift.

Right into the middle of the activity rode the *New Orleans,*

Robert Fulton's first side-wheeler steamboat on the Mississippi River—her maiden voyage only a month and a half before. After mooring to an island the night before the quakes, the crew awoke to find the island had disappeared below the waters of the river. Many fearful citizens thought the steamboat was responsible for the quakes—after all, this new mode of transport was the "devil's contraption," belching black smoke from its central stack.

Slight shakes continued to occur for a while, but none of the aftershocks held the same power as the initial quakes. For some people, the shaking left a mark of horror, pushing them to move elsewhere. For others, life turned back to routine again with just a few rattles and shakes in between.

But it was not over. Contrary to the "normal" aftershocks from this quake-filled time, a violent shaking occurred on January 23, 1812, more than a month after the initial quakes. It rocked the area at around 9:00 A.M. with an estimated magnitude once again greater than 7.0. More people left in despair, some slipping back to the East Coast, others taking their possessions and heading west. In some ways, by this time, the quake had had a profound effect on the United States' migration history, shuffling the midsection's population and spurring many to head west.

There was one more horrific shock—the earthquake that topped them all: on February 7, 1812, the mightiest quake struck the area at 3:45 A.M. at an estimated magnitude close to 8.0. Its tentacles of destruction reached 200 and 300 miles (322 and 483 kilometers) away, in Louisville and Cincinnati, respectively.

The horror on the human side was perceptible in this account (from a letter dated New Madrid, Louisiana [now Missouri], June 13, 1812, and talking about the February 7 quake):

"The country, as you have no doubt heard, has experienced

great injury by the violent concussions of the earthquakes. In some places the soil is much injured owing to the explosions from the bowels of the earth, which have caused the sinking of the ground [liquefaction and subsidence], mostly in the Little Prairie, and in consequence of which we have no doubt that many places that were before dry will now in the times of high freshets be inundated. But the greatest injury the country has sustained is the depopulation by the desertion of the inhabitants. The events of the 7th of February last were dreadfully alarming indeed. About three-fourths of the inhabitants left the country, and a great portion of them have not yet returned, prevented by the apprehension of what may yet befall this devoted country, for the tremors of the earth still continue. We have from five to eight shocks in twenty-four hours, and I would speak within bounds to say we have had ten thousand shocks."

The quaking did eventually rumble to a close, the last quake above magnitude 6.0 occurring on February 11, 1812. By winter's end, the twists and turns from the quakes had left few houses within 250 miles (400 kilometers) of New Madrid undamaged. Marshes drained, uplifted 20 to 30 feet (6 to 9 meters); depressions miles wide and tens of feet deep opened up, the waters rushing in from creeks and rivers. Big Lake in Arkansas wasn't the only resulting lake: the quake created a 100-mile- (161-kilometer-) long, 6-mile- (10-kilometer-) wide, and 50-foot- (15-meter-) deep depression, causing the Mississippi to flow backward for a time to fill in the basin, creating Reelfoot Lake in Tennessee. Saint Francis Lake in Arkansas formed from the quake after dropping 50 feet in elevation; today it is a dry basin called the Saint Francis Sunken Lands.

Though the quakes seemed to end on the 11th, the fear was

still present. Imagine what it would be like to go to bed never knowing when or if another major quake would strike. As Mr. Crist wrote the day after the February 7 quake:

"February 8, 1812: If we do not get away from here the ground is going to eat us alive. We had another one of them earth quakes yesterday and today the ground still shakes at times. We are all about to go crazy—from pain and fright. We can not do anything until we can found our animals or get some more. We have not found enough to pull the wagon." One of Mr. Crist's final entries about the New Madrid quakes stated that he would never go back to Kentucky again.

Accurate Accounting?

But just how many of these accounts are reliable witnesses to the horrendous quakes? The settlement of the central United States was well under way by 1811, and the Mississippi River was becoming notorious for trading goods up and down the length of the country. It was the edge of the western frontier, with a sparse population making communications poor between towns and the big cities along the East Coast. The population was not technologically advanced, and, scientifically speaking, there was a definite lack of earthquake knowledge. Who would have guessed all the shaking was a quake, especially in an area that never experienced such seisms? Legends and myths grew, fueled by fear and superstition. Even illiteracy added to the lack of solid information and interpretation of the events.

The examination of the historical records hasn't been easy. For example, one of the prime accounts of the quake is often thought to be Eliza Bryan's journal. But no one knows if there is such a journal, or if it ever existed. Supposedly Bryan wrote most of the details of the quake from memory—in 1816—that is, if the letter to a Reverend Lorenzo Dow is authentic. And we all know how memory isn't always reliable. In the alleged letter, Bryan recounts several terrifying events that seem accurate, albeit described in less than stellar detail:

"On the 16 of December 1811 about 2 oc. am. [*sic*] we were visited by a violent shock of earthquake accompanied by a very awful noise resembling loud but distant thunder but hoarse and vibrating followed by complete saturation of the atmosphere with sulphurous [*sic*] vapor causing total darkness. The screams of the inhabitants, the cries of fowls and beasts of every species, the falling trees and the roaring of the Mississippi, the current of which was retrograde for a few minutes owing as is supposed to an eruption in its bed, formed a scene truly horrible."

And concerning the mighty Mississippi: "The Mississippi first seemed to recede from its banks and its waters gathered up like a mountain leaving for a moment many boats which were here on their way to New Orleans on the bare sand in which time the poor sailors made their escape from them. Then rising fifteen or twenty feet perpendicularly and expanding as it were at the same time, the banks were overflowed with a retrograde current rapid as a torrent; the boats which before had been left on the sand were now torn from their moorings and suddenly driven up a little creek at the mouth of which they had laid, to a distance in some instances of nearly a quarter of a mile. . . . The river falling immediately as rapidly as it had risen receded within its banks with such violence that it took with it whole groves of young cotton wood trees which had ledged its borders. They were broken off with such regularity in some instances that person [*sic*] who had not witnessed the fact could be with difficulty persuaded that it had

From the New Madrid quake—a huge fissure in the ground. This is one side of a fault trench or "fissure" near the banks of Saint Francis River, Clay County, Arkansas, 1904—years after the quakes, but still visible. *Courtesy of USGS.*

not been the work of art. The river was literally covered with the wrecks of boats."

This is not to say there was no other evidence. There were more scientifically useful accounts, according to a report by Arch C. Johnston at the Center for Earthquake Research and Information at the University of Memphis, Tennessee. The following was a description by a traveler with a more scientific-observation bent who was caught in a huge liquefaction event during the major quakes (much of the New Madrid fault zone lies beneath water-saturated loose soils of the river valley, and the shaking caused extensive liquefaction of the material; this report is probably one of the best accounts of such an event during one of the major quakes):

"I happened to be passing in its neighborhood where the principal shock took place . . . the water that had filled the lower cavities . . . rushed out in all quarters, bringing with it an enormous quantity of carbonized wood . . . which was ejected to the height of from ten to fifteen feet, and fell in a black shower, mixed with sand which its rapid motion had forced along; at the same time, the roaring and whistling produced by the impetuosity of the air escaping from its confinement, seemed to increase the horrible disorder of the trees which everywhere encountered each other, being blow up [sic] cracking and splitting, and failing by the thousands at a time. In the mean time, the surface was sinking and a black liquid was rising up to the belly of my horse, who stood motionless, struck with a panic of terror. . . . These occurrences occupied nearly two minutes; the trees, shaken in their foundation, kept falling here and there, and the whole surface of the country remained covered with holes, which . . . resembled so many craters of volcanics."

Quake Numbers

Overall, amazingly few firsthand accounts were recorded in the three months of quakes—even in regional newspapers—and there are major gaps in scientific accounts of the event. It took a hundred years before Myron L. Fuller of the USGS compiled and published the effects of the great quakes, a time when most of those who'd experienced the shaking were long dead.

Not much else was done until the 1970s, when Dr. Otto W. Nuttli of Saint Louis University finally drew attention to the New Madrid events, the geology of the area—and the potential for strong quakes in the region. Nuttli and his fellow geologists' concerns—and those of others since then—were obvious: the sparsely populated area in 1812 had few if any reported deaths; today, such quakes in this heavily populated area would cause major devastation. They also knew the site had not been silent since the tumultuous year of 1812: besides smaller quakes, a magnitude 6.4 quake was centered near Marked Tree, Arkansas, in 1843; another, of magnitude 6.7 in 1895 near Charleston, Missouri, was felt in 23 states.

Over the past three decades, scientists and historians have continued to gather physical and historical information to support Nuttli's claim. They still rely on the past to present possibilities of the future: the 1811–1812 quakes.

Even though there are still debates about the exact numbers, some reports state that the two largest quakes were greater than magnitude 8.2 on the Richter scale; aftershocks included around two at magnitude 8.0, five at about magnitude 7.7, and ten at about magnitude 5.3—and more than 89 at magnitude 4.3. Another report estimates that the biggest four quakes had body-wave magnitudes between 7.0 and 7.4, surface-wave magnitudes between 7.8 and 8.3, and Modified Mercalli intensities of XI to XII.

During and after the major quakes, an area about 50,000 square miles (129,499 square kilometers) around the New Madrid epicenter was the recipient of Nature's greatest fury. There were no seismographs in place at that time, so scientists have estimated—based on historical and geologic records—the quakes' magnitudes. Just outside the epicenter zone, within a few hundred miles, towns, cities, and villages experienced strong shaking from the major quakes and a Modified Mercalli intensity of about VII—or: "Everyone runs outdoors; damage to building varies depending on quality of construction; noticed by people driving automobiles," or in this case, driving a buggy or riding a horse. Right at the epicenter, the four greatest quakes were about Modified Mercalli intensities of XI to XII—or the two highest intensities. (For a listing of the Modified Mercalli scale, see chapter 4.) In the case of XI, "few structures remain standing; bridges destroyed; broad fissures in ground; pipes broken; landslides; and rails bent"; even more intense, the XII reading is described as "damage total; waves seen on ground surface; lines of sight and level distorted; objects thrown up into the air."

This definition corresponds perfectly with some historical accounts—especially those that mentioned wide fissures in and uplift of the ground, collapse of hillsides and bluffs, lakes forming from river backflow, and liquefaction. The quakes didn't cause devastation only on the surface—the subsurface rock was also affected. Underneath the soil, the largest quakes caused at least six ruptures of intersection fault segments. One of these broke the surface as a thrust fault, causing disruption of the Mississippi River in at least two—some say four—places.

Other places felt the wrath of the New Madrid quakes. The seismic waves of the biggest quakes were felt throughout the eastern United States and Canada—from the Gulf Coast to the Atlantic coast and on to Quebec, an area covering about 2 million square miles (5,179,976 square kilometers). Reports of church bells ringing in Boston 1,000 miles away is a favorite story—but also shows the power of the strongest quakes.

Still another quake description came from Daniel Drake of Cincinnati, Ohio—a state considered to be on the periphery of the New Madrid quake zone. His observations, published in 1815, are one of only a few accounts of what actually occurred during the 1811–1812 quakes. In his book *Natural and Statistical View, or a Picture of Cincinnati* he wrote:

"At 24 minutes past 2 o'clock A.M. mean time, the first shock

From the New Madrid earthquakes of 1811–1812: this landslide trench and ridge in the Chickasaw Bluffs east of Reelfoot Lake, Tennessee, resulted from the New Madrid quakes. *Courtesy of USGS.*

occurred. The motion was a quick oscillation or rocking, by most persons believed to be west and east; by some south and north. Its continuance, taking the average of all the observation I could collect, was six or seven minutes. Several persons assert that it was preceded by a rumbling or rushing noise; but this is denied by others, who were awake at the commencement. It was so violent as to agitate the loose furniture of our rooms; open partition doors that were fastened with falling latches, and throw off the tops of a few chimnies [*sic*] in the vicinity of the town. It seems to have been stronger in the valley of the Ohio, than in adjoining uplands. Many families living on the elevated ridges of Kentucky, not more than 20 miles [32 kilometers] from the river, slept during the shock; which cannot be said, perhaps of any family in town."

It's interesting to note that Drake's record gave scientists a glimpse of what happened in this part of Ohio during the quake. There were no seismograms to take magnitude readings, so scientists have had to estimate the energy from the quakes based on physical evidence. Drake's account showed that the first big quake was probably a Modified Mercalli intensity of VII based on the mention of the chimneys falling. Second, the account points out a major phenomenon during a quake: the violence of an earthquake usually depends on what lies below your feet, even in a small area. Because Cincinnati was built on the loose sediment of the river and not the bedrock "ridges of Kentucky," the quake was felt more—direct evidence that thick, unconsolidated sediments amplify the ground shaking during a quake.

NEW MADRID TODAY
Newest New Madrid

The scientist took a deep breath. He knew the people who still had television or at least radio contact would want to know why they had been shaken so badly—why their world seemed to be coming apart at the seams. Even he was a bit concerned—no matter how many times he checked, his cell phone still read "out of range." He finally accepted the fact that cell phone towers were probably down all over the region. There was no way to find out if his wife and son were all right.

The broadcaster held his hand to his headpiece, then nodded to the scientist. At least they were on the air.

"Dr. Baer is the head seismologist with the local U.S. Geological Survey near Memphis. Can you explain what happened here, Dr. Baer?"

Baer took a deep breath. This wasn't going to be easy. He knew any explanation would seem trivial compared to what was really happening deep underground. "Well, I guess I should start at the beginning..." he said, coughing not only to clear his throat but to search for the right words. "Deep in the layers of rock just below New Madrid is a long, somewhat broken series of cracks in the crust. They no doubt have been there growing, cracking, and straining for millions of years—evidence of how much stress our planetary crust has endured. The—"

Two fire engines drove by, their sirens blaring. Baer could see a flickering light in the distance. Probably another fire set off by a gas-line rupture or transformer explosion. There had been several in the past hour since the Moment Magnitude 7.1 earthquake had struck.

He continued, talking louder as the trucks ran by. "Uh, the New Madrid seismic zone... or the New Madrid fault zone... we call it both... it's in a weak spot known as the Reelfoot rift located deep underground. It's invisible to those of us on top of the crust—there's just no evidence at the surface—it's known mostly in the world of ... well, scientists like myself. But trust me, it's there, running in a crazy path that takes it 150 miles southward from Cairo, Illinois, through New Madrid, Missouri, down through Blytheville, Arkansas, dipping into Kentucky near Fulton, and southeast to Dyersburg, Tennessee. Not only does it pass through five states, but it crisscrosses the Mississippi River at least three times."

Baer watched as the interviewer turned white at his last remarks. That's when, too, they all felt the violent aftershock.

—*Fictitious account of what could happen if the same types of quakes that occurred in 1811 to 1812 struck the area of New Madrid today*

Yes, the above is fictitious—but the New Madrid fault zone and all the details related by our "Dr. Baer" are not. In fact, this "crack" in the Earth's crust has the potential to become one of the most deadly earthquake centers North America has ever known.

There's good reason why scientists are nervous about the New

Madrid fault system and its potential to shake again. Paleoseismologists, scientists who study ancient evidence of seismic activity, have uncovered some disturbing evidence of earthquakes in the same region—and not only the ones in 1811 and 1812. What has been found is alarming: in the 2,000 years before the nineteenth-century quakes, this area may have had as few as two or as many as four major earthquakes. One was just north of New Madrid and likely occurred between 800 and 1000 AD; another is closer to Marked Tree; and possibly another one occurred near Blytheville—both in Arkansas—thought to have shaken the area between 1200 and 1400 AD.

Again, since recorded history of this area began a mere 200 years ago, scientists can only go by physical evidence in the field. And who knows what evidence could have already eroded away, or even been dug up for building infrastructures, commercial buildings, or housing?

Take a good look at the sinuous, wandering Mississippi River, which makes up some of the borders of Tennessee, Missouri, and Arkansas. Go to the far northwest corner of Tennessee—an area so close to New Madrid, Missouri, you could practically toss a coin across the river. Now check out the far southwest corner of Tennessee about 130 miles (209 kilometers) away. There sits Memphis, Tennessee, home of Graceland, numerous colleges, and a population of about 680,000; the greater Memphis metropolitan area takes in about 1.25 million people—the second largest in the state only to Nashville. It is also a city within reach of the New Madrid fault zone—and scientists know quakes similar to those in 1811 to 1812 would almost annihilate the entire area.

The Future of Quakes along the New Madrid

In Los Angeles, students practice citywide earthquake-preparedness drills to test whether they're ready for "the big one." Most schools stage them twice a month just to stay in practice. In some areas, like Laurel Canyon, south of Los Angeles, some schools also require that every child has an individual one-gallon resealable bag of necessities including water, nonperishable foods, a disposable flashlight, and a comforting note from a parent. Teams of teachers have also been trained in search and rescue and firefighting in case of a major event. And each school must keep a three-day supply of food and water for each student; keep vigilant about insecure drawers, doors, and boxes; and finally, train all students as to where to hide in case of a quake.

Will this—or should this—be the practice of children who go to school in the New Madrid fault zone?

Today, the New Madrid seismic zone is often rocked by moderate to minor quakes—an average of about 150 per year. The largest felt in recent times happened in 1976—a magnitude 5.0 with a 4.5 aftershock—and on September 26, 1990—a magnitude 4.8. According to the scientists at the Center for Earthquake Research Information, as of 2000, the probabilities for earthquakes at various magnitudes are as follows:

Magnitude	Probability in 15 years	Probability in 50 years
6.3	40-63 percent	86-97 percent
7.6	5.4-8.7 percent	19-29 percent
8.3	0.3-1 percent	2.7-4.0 percent

Of course, there are other ways of looking at this data. Here is another chart provided by the CERI—magnitude and the expected quake rate for the central United States versus the world:

Magnitude	Central United States	The World
3.0	6 per year	250 per day
4.0	14 months	40 per day
5.0	10-12 months	8 per day
6.0	70-90 years	2 per week
7.0	250-500 years	20 per year
8.0	550-1,200 years	2 per year

All of these estimates are based on previous data—and those "previous data" are scarce. Because there are so few large quakes, or even moderate quakes, it is difficult to determine the actual probabilities. But again, statistics are merely a grouping of numbers that predict a possible pattern. And even though some scientists believe they have detected a pattern to such moving faults as the San Andreas, sometimes there is no pattern.

So if there is only a small chance of a major quake around New Madrid, why are scientists constantly warning the public about a potential earthquake in the area? The biggest reason has to do with experience: recent earthquakes have shown that even moderate shaking can cause loss of life and major destruction of property. For example, the magnitude 6.9 earthquake in Kobe, Japan, killed 5,500 people and caused $100 billion in damages. This earthquake was mostly due to the liquefaction of loose soils that underlie the Kobe wharves, causing the ground to become unstable. In the Mississippi Valley, a moderate quake occurs more frequently than a quake of magnitude 7.0 or 8.0. But even a moderate quake could cause great damage, as the loose valley soils would amplify the energy of the seismic waves.

Of course, there are other earthquake-occurrence estimates on par with the chart above. For example, a 2000 study from the University of Colorado at Boulder suggested that the potential for a large earthquake along the New Madrid seismic zone in the central Mississippi Valley was "serious." Still another study states that the probability of having a magnitude 6.0 to 7.0 quake within the next 50 years in the New Madrid region is greater than 90 percent. If true, such a quake would be devastating, as most structures are not built for such quakes—and because the region's modern cities have such large populations.

As scientists—and the public—have discovered over the years, these statistics are merely estimates. And often estimates are wrong. Just ask your local weatherman when it comes to forecasting your weather—events that occur *above* ground.

Spreading from New Madrid

Earthquakes that occur thanks to the New Madrid quake zone don't happen to people just around New Madrid, Missouri. For nearby states, the quake-out-of-nowhere scenario isn't too far-fetched. There is good reason to be concerned: remember, earthquakes in the central or eastern United States—like the New Madrid—affect a much larger area than earthquakes of similar magnitude in the western states. This has to do with the rock and soil—the more solid the crust under your feet, the less shaking; the more loose soil, the more the energy is propagated. For example, the San Francisco magnitude 7.8 earthquake of 1906 was felt a mere 350 miles (563 kilometers) away, in the middle of Nevada; as previously mentioned, the larger New Madrid quake of 1811 rang Boston church bells 1,000 miles (1,609 kilometers) away. So if there is a possibility of a quake along the New Madrid, the surrounding states should pay attention.

There are some actions being taken. For example, when Arkansans opened their newspapers in late May 2006, they were in for a surprise: federal and state emergency officials announced that it was time for all the state's residents to prepare for an earthquake—now. The warning was from the Federal Emergency Management Agency (FEMA) at Little Rock, issuing a reminder to residents that catastrophic earthquakes are unpredictable.

What was the reason for the warning? No, there was no dire prediction from a psychic. The reason was more scientific: the New Madrid fault zone—a fault that runs from Marked Tree, Arkansas, north-northeast up through the boot heel of Missouri, to near Cairo, Illinois, is right in their backyard. And of all the spots in the United States not near an earthquake-prone area, this area of the Midwest has the greatest potential to shake more than one state. Not only that, since the rocking of the region between 1811 and 1812, Marked Tree is one of the places that has had a major quake—magnitude 6.5 in 1843—that caused major damage.

Another was on October 31, 1895—a magnitude 6.8 near Charleston, Missouri. The so-called Halloween quake was felt as far away as Pittsburgh, New Orleans, and Topeka. It was felt in a wide circle encompassing several states—many of the people fearing that another succession of 1811–1812–type quakes was about to arrive. For example, in Chicago, windows rattled and the 12,000 telephone switches on the main Chicago exchange lit up simultaneously, causing chaos for the phone company. People raced into the street in their nightclothes, and old brick buildings cracked in Saint Louis, Missouri. Thousands were shaken awake in Indianapolis; electric lights flickered and windows rocked in Kansas City, Missouri. The worst damage was in Cairo and Charleston: hundreds of chimneys broken, plate-glass windows, china, and glassware shattered—and every building in Charleston's commercial block experienced some damage.

The Halloween quake was similar in strength to the 1994 Northridge, California, quake (magnitude 6.7) that resulted in about 33 deaths and $20 billion in damages. But the Missouri quake didn't cause as much damage, and there are no records of any deaths. The biggest factor, of course, was the lack of population and structures—this area of the central states was mostly desolate around the late nineteenth century. A similar strike today would lead to substantial losses—both in terms of structures and lives—not only because of an increase in population, but as mentioned before, because of the reaction of the surface. In this region, the strong mid-continental crust radiates the damaging seismic waves much farther than the continental crust in California.

What about another neighboring state—Ohio? By 1811, the state had a population of just over 230,000. And even though moderate quakes had been felt in the Ohio Valley in 1776 and 1791 (or 1792), they were apparently not even close to the shaking that started on December 16, 1811. The P-waves hit Cincinnati only a minute and 18 seconds after the initial quake; then came the slower surface (S) waves, the rocking reportedly lasting for several minutes. It was the first of four major shocks the region would feel, along with hundreds of small ones, over the next few months.

What about Ohio—a state "merely" on the periphery of the New Madrid quake zone—today? The Ohio geological survey published its own estimate in 1998, stating that "the probability of a recurrence of the 1811–1812 [New Madrid] sequence is low for the immediate future, perhaps only 10 percent or less in the next 50 years." They also guesstimate that earthquakes of such magnitude—based on paleoseismic data—occur in the New Madrid region only once every 500 to 600 years.

But they are quick to point out that such estimates are merely estimates—that nature doesn't always follow a regular pattern. In addition, even predictions of possible quakes vary greatly depending on the computer model. Data seem to suggest that a magnitude 8.0 earthquake event in the next 50 years has a probability of 2.7 to 11 percent; the probability of a 6.0 to 6.5 event in the next 50 years raises the ante, but is also a considerable spread—between 45 and 97 percent.

Perhaps what many scientists are concerned about are the earthquakes measuring around the magnitude 6.5 range, similar to the 1895 Charleston, Missouri, quake. Based on the historical evidence, such quakes appear to occur about once every century in the New Madrid zone. Since one is definitely overdue, many scientists believe there is a high probability that such an event will occur in the near future.

What's the Cause of It All?

According to some reports, the 1811–1812 New Madrid earthquakes may be a continuation in a series of quakes that occurred in 1699, 1776, 1792, 1795, and 1804. No one truly knows if these predecessor quakes were stronger than the New Madrid quakes, or if some of the noted changes in the 1811–1812 New Madrid earthquake-induced features actually occurred earlier. But series of quakes or not, what do scientists believe causes this shifting around the New Madrid area?

The reason, again, is underground: some scientists believe the numerous faults underlying this area of the Midwest lie in the intruded crust of a failed intracontinental rift called the Reelfoot rift. Here, the crust tried to establish a rift valley—to pull apart and slowly change the continental landscape similar to how today's Great Rift Valley is tearing apart eastern Africa. Though it didn't create a rift, the overall fault system is still active, generating an average of 200 earthquakes a year, with only about 8 to 10 actually large enough to be felt. And it's the active faults and continual quakes—and evidence of the past horrific quakes in the region—that deeply concern scientists who study today's New Madrid fault zone.

There is another explanation, which could be another or an additional cause: according to studies conducted by geophysicist Mark Zoback at Stanford University, the reason why the New Madrid fault zone shakes is because of an ice-age legacy. Tens of thousands of years ago, glaciers covered the northern part of North America, extending down into what is now the central United States—more specifically, as far south as central Illinois. The weight of the massive miles-high ice sheet pushed down the lands underneath and just beyond it. Though the ice never reached the now-known New Madrid fault-zone area, its crushing effects made an impact. But it wasn't over: as the ice sheet melted and retreated, the land rebounded, releasing pressure that translated into earthquakes. The shaking hasn't stopped yet, and according to Zoback, quakes will probably persist for thousands of years more as the land tries to reach equilibrium.

The concept of a rising land after the retreat of the continental ice sheet is not new. In New York's Adirondack Mountains, there are plenty of earthquakes—and the area is nowhere near a plate boundary. The region's rocks are around a billion years old; they were uplifted a mere 5 million years ago. The largest of the quakes so far measure above Moment Magnitude M_w 5.0, but as to why, there is still speculation. Some theories point to a possible hotspot below this part of the North American continent; another theory states that it was in an ancient chain of active volcanoes that extends into New England. Still others speculate that, like the New Madrid fault zone, the ice age ice sheets were responsible: the mountains were not only scraped down by the massive Laurentide ice sheets, but retreating glaciers also released pressure on the rock. Thus, the Adirondacks, too, probably experience quakes as the ground slowly rebounds.

CHAPTER 7

More Cases of Rogue Earthquakes

The day had been an exceedingly hot one and the evening was unusually sultry, with such a profound stillness in the air that it provoked general remark. . . . The temblor came lightly with a gentle vibration of the houses as when a cat trots across the floor; but a very few seconds of this and it began to come in sharp jolts and shocks which grew momentarily more violent until buildings were shaken as toys. . . . No need to tell of the horrors of that moment or of those succeeding. The fact that lighter shocks continued at frequent intervals throughout the long, dreary night kept the nerves of all keyed to such a high tension that it is not strange that several persons lost their reason. . . . It was not until the next day that the people began to realize the extent of the calamity that had befallen them. Then it was learned that not a building in the city had escaped injury in greater or less degree. Those of brick and stone suffered most. Many were down, more were roofless, or the walls had fallen out, all chimneys gone, much crockery, plaster, and furniture destroyed. . . . It seemed on the first survey that all public buildings and the principal business blocks were utter ruins. Many of them had to be torn down and were rebuilt.

—*A partial account of the great earthquake of 1886 that destroyed the city of Charleston, South Carolina, by Paul Pinckney,* San Francisco Chronicle, *May 6, 1906*

Along with being one of the major financial centers of our global economy, Hong Kong has some of the tallest, most closely packed high-rise buildings on the planet. On September 14, 2006, at about 7:53 P.M. local time, a rare earthquake measuring magnitude 3.5 shook the densely populated city (there are around 7 million people living there); the tremor lasting a few seconds. Though the Hong Kong Observatory recorded the epicenter to be 22 miles (36 kilometers) south-southwest of the city, residents reported scary results from the mild quake. Walls and furniture started to shake; one woman said her entire house began to move; and residents began wandering the streets, wondering if they were being attacked by terrorists, or if there had been a gas line explosion or even a plane crash. This was not typical for Hong Kong, a city that, since 1979 when the observatory opened, has experienced only 51 tremors measuring up to magnitude 4.0. And since 1874, only six tremors measuring over magnitude 5.0 have been felt in the region. This was essentially a rogue earthquake—one of many that occur in the wrong places every year.

Until the past decade, little was known about faults and their possible respective quakes under the intraplate regions of the world. Now scientists are studying these fractures with more diligence—especially since they realize that some of the faults may have the potential to shake hard, not only in sparsely populated areas, but under some of the largest cities in the world. They don't know all the answers—but they know many of these places experience quakes far from the borders of a tectonic plate edge.

Just consult the earthquake hazards map. The possibilities seem endless and scary to those of us who see a bull's-eye drawn on the map—right where we live.

The following are a few of those instances where earthquakes have occurred in the wrong places. In most cases, scientists know what regions have shaken in the past—which only makes them more cautious about future quakes.

MORE CENTRAL STATES QUAKES
Close, but Not from the New Madrid
Though the New Madrid fault zone gets most of the publicity (for obvious historical reasons), there are other fault zones in the central United States. Some of these regions did feel the effects of the New Madrid 1811–1812 quakes, and if there was another large quake near New Madrid, they would no doubt feel the shaking again. But scientists are now interested in other places, where larger fault systems cut through the central states—alarmingly, many of the faults with the potential to produce their own quakes.

Take a system relatively close by: the New Madrid area butts up against the Wabash Valley fault system (often called the Lower Wabash Valley fault system) to the north—a valley that borders both southeastern Illinois and southwestern Indiana. Recent studies of this fringing fault system show that the area may have experienced major earthquakes in the past. The Wabash system is a series of north-northeast-trending normal faults that have been easily mapped at the surface. It also includes a basin that contains, in some spots, more than 12,000 feet (366 meters) of sediment accumulation—loose material that shakes quite nicely during a moderate to major earthquake.

Like the New Madrid fault zone, scientists believe that the Wabash system began as a rift that failed to open during a continental breakup hundreds of millions of years ago. The development was typical of features called *sag basins* that subside and fill with sediments—usually shallow water deposits of mud, silt, and sand on the periphery, and marine carbonates at its deeper parts. As the continents shifted and moved, the area was subjected to alternate periods of compression and tension, thus forming the active and dormant faults of the region.

The reason for the present concern about the Wabash Valley is the past: prehistoric quake magnitudes probably measured on the order of Moment Magnitude M_W 7.1 and 7.5—a number that outshakes the region's largest historical-record quake of Moment Magnitude M_W 5.5, which occurred on November 9, 1968. Scientists have based the ancient higher quake magnitudes on liquefaction features—more than 1,000 paleoliquefaction dikes, also known as *sand blows*. When strong quakes release energy, the shaking can cause layers of water-saturated sandy soil—typical for the somewhat loose sediment in this basin—to behave like fluid under pressure. Like volcanoes in miniature, the pressures can cause the liquefied sand to move through cracks and crevasses in the overlying soil, forming sand blows.

The Wabash dikes—almost all exposed along banks of rivers—are typically filled with sand and gravel. They lie nearly vertical, with some extending up to almost 15 feet (4 meters) and with widths ranging from 3 feet (2.5 meters) to 12 inches (30 centimeters). Based on radiocarbon dating of the sand blows and even artifacts of people once living at certain sites, the largest event,

From the Seattle, Washington, earthquake of February 28, 2001: this store received extensive damage during the quake. There was also substantial damage in Pioneer Square, part of Seattle's historic district. *Courtesy of FEMA.*

measuring magnitude 7.5, probably occurred about 6,100 years ago. This major quake—even more powerful than the one that struck Los Angeles in January 1994—probably had an epicenter near Vincennes. Further study of the sand blows determined that there were at least six major quakes with epicenters in Indiana during the past 12,000 years—a short period in geologic time.

But it's not only the past that concerns scientists who study the Wabash system—it's the present. Just why do quakes continue to occur in this wrong place so far from plate boundaries and volcanoes?

And there is a troubling theory: some scientists believe the Wabash Valley fault system is no longer active, and that more recent quakes are possibly related to breaks in the crust elsewhere.

But just where is a matter of speculation. Scientists who study this area northeast of New Madrid have the same problem as those who study the New Madrid fault zone. The infrequency of major earthquakes makes it impossible to understand the potential, place, or possibility of the next damaging quake.

There is some quake data, albeit not much. More recent quakes have ranged from small trembles to decent shakings. On June 10, 1987, a Moment Magnitude M_W 5.1 quake occurred near Lawrenceville, Illinois; and on June 18, 2002, a temblor of Moment Magnitude M_W 4.8 shook the Greater Evansville area, with an epicenter not 10 miles (16 kilometers) from the city. Yet another quake occurred in November 1968, measuring Moment Magnitude M_W 5.5; but this quake's epicenter was west of the Wabash Valley fault system in south-central Illinois (often included on the list because it caused damage in the region). These quakes caused minor damage, such as bricks thrown from chimneys and cracking of older masonry and stucco-type buildings. But it's still not known if the Wabash system caused the quakes.

SOUTHEAST QUAKES
Charleston, South Carolina, the Largest Yet
The 1811–1812 New Madrid earthquakes were not the only ones in "unlikely" places, or even the largest in the eastern United States. The 1886 Charleston, South Carolina, earthquake holds that distinction; it was the highest-magnitude quake ever to hit the East in its 200-year-plus recorded history. Estimates put the quake at around Moment Magnitude M_W 7.3—and thanks to hitting a populated and developed part of the coast, it is considered the most (so far) damaging earthquake ever to strike the

The Charleston, South Carolina, quake of 1886 caused a craterlet that formed on Ten Mile Hill. These are actually sand blows, or sand volcanoes, of the craterlet type, found in the area of major shaking. These craters are often filled in by sediment years afterward, making them difficult to discern unless they are found right after a quake. *Courtesy of USGS.*

southeastern United States. And it's nowhere near a tectonic plate boundary.

This time, nature seemed to fire more than a few warning shots: on August 27, 1886, in Summerville, a small town about 25 miles (40 kilometers) north of Charleston, something seemed to lightly shake a few restless sleepers from their beds at about 1:30 A.M. local time. By 8:30 A.M., those who didn't feel or hear the nighttime shaking heard an explosion. Speculations included some darn fool blowing up gunpowder, a train boiler, or the nearby phosphorus plant exploding.

The real explanation wasn't possible—or even thought about.

Most people, except some longtime residents, had never even heard of an earthquake, much less felt one. And even when a Charleston resident telegraphed the news to the *Atlanta Constitution*, the editors laughed, probably thinking the heat was finally getting to their neighbors to the north. The next day, in reference to the shaking, the paper stated that it was "merely imagination ... simply a shock accompanying the announcement that Atlanta had won yesterday's game from Memphis, ensuring it the pennant." Baseball obviously overruled a possible earthquake in the south.

Summerville wasn't lucky enough to get away with just two rumbles. At 1:45 P.M. on August 28, an even stronger shock threw pottery around, broke windows, and caused a general uneasiness. After all, few people knew what the shaking portended—there were few people old enough to remember the New Madrid shaking of 1811–1812, and most of the damaging quakes were centered around Missouri and the distant Mississippi River. Aftershocks kicked in that day and the next—most just tremors, but some felt as far away as Wilmington, North Carolina.

On the last day of August, the real activity began at around 9:50 P.M., but this time, it was down the road in Charleston. The quake lasted just under 60 seconds, but succeeded in damaging almost 90 percent of the city's brick structures. Other structures didn't do well, either—whether commercial, residential, brick, or wood, on high or low ground, filled-in sediment or not. More than 2,000 buildings were destroyed, and every structure in the city had some type of damage.

The shaking also caused a multitude of fires, damaged wells, tore apart water and sewer lines, bent railroad tracks (and even tossed one train off the tracks), and cracked a dam in Langley, sending floodwater to the town below. It's estimated that 110 people lost their lives, but that, too, is only a guess—at that time there was no census to keep track of people. In 1886 dollars, the damage in Charleston alone totaled about $6 million; in today's dollars, the cost would be in the billions.

Though it was difficult to prove, a scientist reported two epicenters (in 1889) responsible for the quakes: Woodstock, a railroad stop on the Southern Railway leading into Charleston and 21 miles (34 kilometers) northwest of the city; and Ravenel, a small town 23 miles (37 kilometers) southwest of Charleston. The physical consequences were typical for a major quake, especially the liquefaction of loose coastal sediments that created sand blows and sand craters, some measuring more than 21 feet (6.4 meters) wide. There were no surface ruptures, though several fissures were reported parallel to streams and canals near the epicenter.

The shaking spread out over 2.5 million square miles (6.5 million square kilometers) of land from the epicenter—stretching to cardinal points from Cuba to the south; New York (some reports say Boston) to the north; Bermuda to the east; and the Mississippi River and New Orleans to the west. It was strong enough to sway those ubiquitous church bells that long ago always seemed to ring during quakes—this time, down in Saint Augustine, Florida. It shook towns along most of the eastern coast, with reports of cracked walls in Harlem, New York. Cities within about 200 miles (322 kilometers) were hardest hit, with housing and structural damage. They included some of today's largest southern cities: Columbia, South Carolina; Augusta, Georgia; Raleigh, North Carolina; and Atlanta and Savannah, Georgia.

Because most faults and other geologic structures related to quakes are hidden under a thick layer of coastal sediments, it's difficult to discern the culprits "at fault" for the Charleston quake and other quakes in the region. One suggestion is Cameron's Line, a fault that runs through southeastern New York and follows a crooked line that runs all the way down to Charleston. Still other faults around the Cameron may have been the reason for the shaking, especially the north-northeast-trending Woodstock fault and the northwest-trending Ashley River fault. Scientists speculate that these faults are due to a "rift" in the crust—a leftover from the separation of the Europe and Africa plates from the North American plate 400 million years ago. This "parting of the ways," when the great continent of Pangaea split apart to create the Atlantic Ocean, also created a legacy of faults thought to be responsible for the 1886 quakes and others since.

For around 35 years after the massive Charleston quake, some 300 aftershocks were felt—four of them greater than magnitude 5.0. The quakes have a pattern, too—about 70 percent of the quakes along the state's coastal plain have occurred in the highest-intensity area of the 1886 earthquake, with the majority clustering around three areas west and north of Charleston: Ravenel–Adams Run–Hollywood, Middleton Place–Summerville, and Bowman. Even today, South Carolina continues to have quakes, averaging about 10 per year and ranging from less than magnitude 1.0 to about 3.8 on the Richter scale, with the strongest recent shaking a magnitude 4.1, on August 21, 1992.

The Other Carolina
North Carolina is divided into three regions—the Blue Ridge, Piedmont (foothills), and the Coastal Plain—along with a collection of geologic structures that have given the state its share of faults. Two of the main cracks are the Jonesboro fault, which runs in a southwest to northeast direction from the South Carolina

line, between Raleigh and Durham, to near Virginia; and the Nutbush Creek fault, which runs north of Raleigh to the Virginia line. Both are seemingly stable, but like all faults—even those in the middle of a tectonic plate—they can move ever so slightly, causing earthquakes.

So far, most recorded shakings in the state have been between magnitude 2.0 and 4.0. But that doesn't mean they have been exempt from major shakes. The earliest North Carolina earthquake on record was from March 8, 1735, near Bath. Several others occurred in succession in January into the summer of 1874, when the Bald Mountain region shook, rattling plates on pantry shelves and breaking windows. As the shaking seemed to increase in intensity over those months, several residents packed up and left—afraid the mountain was actually going to blow, or at least chase them with streams of lava.

The largest quake recorded so far in the area occurred in the Smoky Mountains region, a magnitude 5.2 near Waynesville on February 21, 1916. During this quake, there were enough people in both North Carolina and neighboring Tennessee to really notice the shaking. Chimneys fell to the ground, windows broke in many houses, and people rushed into the streets in a panic. About 43 miles (70 kilometers) west of Waynesville, the flow of water in springs increased, and in some areas, became muddied after the earthquake. It also rattled homes and people in about a 200-square-mile (518-square-kilometer) area—into Alabama, Georgia, Kentucky, Virginia, and South Carolina.

But the really damaging quakes have come from outside the state's borders—from South Carolina to Missouri. For example, the aforementioned Charleston, South Carolina, quake's great magnitude was amplified by the soils and rock of the region, sending seismic waves out in all directions. Right next door, of course, was North Carolina. In one report, residents in Asheville, apparently still lounging around at ten in the morning, were shaken out of their beds; in another, houses cracked, and some nearer to South Carolina collapsed.

The Truth about Seneca Guns

Along the North Carolina coast there's another phenomenon once attributed to earthquakes—the so-called Seneca guns. The name is thought to have originated from the James Fenimore Cooper short story "The Lake Gun," mainly because the same types of booming events occur in two of New York's Finger Lakes, Seneca and Cayuga. These *mistpouffers* (they also occur in other parts of the world, and thus this name from the Netherlands and Belgium booms) rock an area like the explosion of a cannon or sonic boom. In North Carolina, the noises have been reported since before the nineteenth century. Houses shake, windows rattle—sometimes many times a month or a few times a year. And some longtime residents say never at night and never on a Sunday.

Thanks to all the seismometers keeping track of crustal movements throughout North Carolina and nearby states, scientists now know the booms *do not* come from earthquakes, storms, or airplanes. And though no one has the definitive answer, there are plenty of guesses: methane explosions on the ocean floor (which doesn't explain this phenomenon in lakes like Seneca, as there is no evidence of methane blobs at the surface); chunks of the continental shelf sliding into the ocean or landslides in lakes; and in North Carolina, even the sudden rush of cold air hitting the Gulf Stream current.

Watch Out, Sunshine State

On September 10, 2006, at 10:56 A.M., a magnitude 6.0 earthquake occurred, lasting for about 20 seconds, complete with a rumbling that shook people up, literally and figuratively. This powerful quake was not along a plate boundary. It occurred in the Gulf of Mexico, in a spot just off Florida's southwestern continental shelf—around 25 degrees above the equator, and about 260 miles (418 kilometers) west-southwest of Clearwater, Florida. The shock waves were felt along the U.S. Gulf Coast and above: Alabama, Kentucky, Louisiana, Mississippi, South Carolina, Tennessee, North Carolina, Georgia, and Texas felt the tremors, as did Freeport in the Bahamas, and Cancún. It was the largest quake to hit the region in three decades, knocking items off shelves and rattling windows.

There was also a sigh of relief. The devastating tsunami of December 2004 in the Indian Ocean region was still fresh in everyone's mind. The only "waves" were seiches in several Florida swimming pools—thankfully, not a huge wave that often follows deep ocean quakes. The energy to create a tsunami needs a plate-boundary shift, and the shallow waters of the Gulf are not conducive to such wave development.

Scientists immediately realized this "rogue" earthquake was not so unfamiliar. There had been another quake, measuring magnitude 5.2, in February 2005, near the same area. Few people noticed that quake, but it was the second largest recorded in the Gulf of Mexico. In fact, in the past three decades, the USGS has recorded more than a dozen moderate quakes in the eastern part of the gulf.

No, this is not the start of an unprecedented trend. Depending on the scientist, there are two distinct possibilities for this rare earthquake: one theory is that sediment that drops into the gulf—from rivers and underwater channels along the coast—deposits enough material to create excessive weight. This weight pushes down the sediment underneath, causing it to shift and produce earthquakes.

The more accepted theory is that a movement along a plate boundary, say in the Caribbean, triggers a "trickle-down" effect, causing this part of the North American plate to move along a weak spot, usually a fault line. And there are plenty of faults in the area, some cutting along the Gulf of Mexico and well inland into Mexico, south and east Texas, and some through Louisiana, Mississippi, Alabama, and the extreme western Florida panhandle.

NEW YORK QUAKES
Quake New York, Shake New York
More than 400 earthquakes have occurred in the Empire State since around 1730. The largest so far occurred in the Cornwall-Massena area (near the United States–Canadian border) on September 5, 1944—measuring a magnitude 5.8 to 6.0 (depending on which report you read) and felt from Canada south to Maryland, and from Maine west to Indiana. In fact, New York ranks as the third-highest earthquake activity level east of the Mississippi River, outranked only by South Carolina and Tennessee. And it, too, is far from a tectonic plate boundary—surely we're talking here about quakes in the wrong places.

Most of the early New York quakes can only be judged based on sporadic reports from the then sparse population. The "quite severe" 1867 quake in northern New York literally shook people out of bed; the 1877 quake was most severe along the Saint Lawrence River and Lake Champlain, causing chimneys to fall, crockery to overturn, and ceilings to crack. A "strong quake," estimated to be about magnitude 5.5, on February 10, 1914, occurred near Lanark, Ontario. Its effects reached out in many directions, including breaking water pipes in Saint Lawrence County; more than a hundred miles away, it caused a cave-in around Binghamton, New York, and cracked streets in nearby Johnson City.

On August 12, 1929, another quake—this one measuring about magnitude 4.4—shook eastern Attica and the region to the east. In the town itself, 250 chimneys tumbled down, several brick buildings were damaged, almost every monument in a nearby cemetery fell, several wells went dry, and a crack formed in the railroad embankment near the railroad station.

From the Charleston, South Carolina, earthquake on August 31, 1886: a displaced coping on the portico of the old guardhouse on the southwest corner of Meeting and Broad streets. Photo by C. C. Jones, September 1886. *Courtesy of USGS.*

And quakes in New York are not just something we hear about in history class. The most powerful quake to hit Long Island since 1937 occurred on October 22, 1981, measuring magnitude 3.5. Its epicenter was in the Long Island Sound, and though it did no major damage, it cracked plaster and rattled dishes and windows. On October 7, 1983, a magnitude 4.89 quake struck the Blue Mountain Lake area in the Adirondacks—a place notorious for its occasional quakes—this one was violent enough to crack chimneys, break dishes, and overturn objects in the area. And a magnitude 4.0 quake was definitely felt on Long Island on October 24, 1985; it was the largest to hit the region in 48 years, with an epicenter in northern Westchester County, in Ardsley.

Fast-forward to 1991, when a quake centered around 40 miles (64 kilometers) west of Albany shook New York and parts of New England—measuring around magnitude 4.0. Still another seism about 15 miles (24 kilometers) southwest of Plattsburgh, New York, was felt on April 20, 2002. It measured a hefty magnitude 5.1, collapsed at least two roads, rattled and broke windows, and cracked foundations. It was felt from Cleveland, Ohio, to Maine and Baltimore, Maryland—and even in New York City.

More recent studies have even added the Finger Lakes region to the list. Research conducted at the University of Buffalo shows that the area has gained a reputation for having a much greater seismic potential than was previously thought. Using a variety of field and satellite data, researchers discovered hundreds of faults throughout the Appalachian Plateau—with some cracks known to have been seismically active over the past billion years. It's a bittersweet situation: scientists know such faults can lead to the discovery of reservoirs for oil and gas; while in contrast, these faults could also mean the potential for a moderate to major earthquake. Most troubling are the faults that may intersect waste dumps, such as the low-level radioactive waste facility at West Valley. But as with most places in the United States in which potentially active faults lie, New York's short historical record, only about 200 years, makes it difficult to draw conclusions about the likelihood of faults having the potential to shake the state.

Quakes in the Big Apple?

New York City seems like the perfect place to keep away from natural hazards. Yes, there's the occasional hurricane that rushes up the coast, possibly sending a storm surge in the direction of the harbor. But there don't seem to be any other "hazards," such as volcanoes, sinkholes, or earthquakes—or are there? In fact, there's one hazard New York City really can't escape—the earthquake. The reason is obvious: Manhattan Island is crisscrossed by earthquake faults.

These faults mean that New York City is not immune to shaking. In fact, the two most serious quakes so far struck in 1737 and 1884, both measuring an estimated magnitude 5.0, respectable for an earthquake. Both quakes had their epicenters in roughly the same spots, in Rockaway—and some seismologists believe that the chances of another Rockaway quake increase if you count the two quakes that have occurred so far. Though these were moderate quakes, if they occurred today they would pretty much decimate the aging housing stock, especially around the Upper West Side.

There have been plenty of others. The year 2001 seemed like the year of the quake: one earthquake occurred on January 17, 2001, measuring 2.4; that may not seem like much—but the epicenter was in midtown Manhattan. Another magnitude 2.9 struck the next month, but the hardly felt quake's epicenter was about 75 miles (121 kilometers) south of Buffalo. The October 2001 quake measured magnitude 2.6, shaking parts of Manhattan and Queens but with no injuries and little damage.

The propensity toward shaking lies underground, and the area has more than one fault running through it. There's a fault between 42nd Street and 34th Street—and it's the same fault responsible for the 1886 Charleston, South Carolina, quake, and for other quake events in the New York City region: Cameron's Line. This fault runs through the Bronx, down the East River, past Gracie Mansion, through Staten Island—and all the way down to Charleston, South Carolina.

But it's not alone: the 125th Street fault zone runs southeasterly from 125th Street (near the Hudson River on the west side), through the northeastern corner of Central Park (near 110th Street and 5th Avenue), meets Cameron's Line at Gracie Mansion, and to East 96th Street near the East River, and along the river. Other faults run across Dyckman Street to the north, along an east-west line that stretches from Broadway in the 200s through Harlem to East 125th Street; and east-west from the Intrepid Museum to Alphabet City. North-south-running faults include one down the Hudson from 42nd Street and into the harbor; and another along Broadway from the upper 30s, through Chambers Street, and on to Battery Park.

The explanation for this series of faults lies in the past. New York City is located almost right where the North American continental plate ripped away from the European and African continental plates. (You can see the famous "jigsaw puzzle" fit of the continents on world maps.) The processes to produce this rift zone took time: about 400 million years ago, Africa slowly collided with North America, consuming a proto–Atlantic Ocean that scientists call the Iapetus Ocean. As it closed, it left behind the long, sinuous, suturelike fault—the geologic feature called Cameron's Line.

Other faults formed as a reaction to the pounding together and tearing apart of the continental plates; this scenario didn't occur just once, but rather several times over hundreds of millions of years. Even the weight of the ice age ice sheets—the ones that advanced and retreated over New York several times and finally retreated a mere 10,000 years ago—probably added some additional cracks in the crust. Over about 2 million years, these glaciers carved valleys in the fault zones, including one visible today—the 125th Street fault zone.

The worst damage from the August 31, 1886, Charleston, South Carolina, quake occurred in the center of the town. *Courtesy of USGS.*

Even faults that don't underlie the city unnerve some scientists. One of particular concern focuses on the Newark Basin, a depression that sits between two belts of fault lines. The first belt runs along the Hudson Highlands. This includes the Ramapo fault system, a fracture about 70 miles (113 kilometers) long, running from Westchester County to the Hudson River. It is also one of the oldest faults—around a billion years old—with activity recorded along the fault since the 1970s. (It is also the reason New Yorkers can live in the city—its "height" and the Ramapo Mountains preventing the sea from filling into the Newark Basin.)

The second belt to the east encompasses parts of Connecticut, Westchester County, and Long Island, down through New York City, and into central New Jersey. The two historical quakes measuring magnitude 5.0 shook along the eastern belt. And over the past quarter century, this basin has produced clusters of quakes—most temblors too small to feel, in the magnitude 1.0 to 3.0 range. Are these seemingly lesser crustal movements precursors to a future moderate or major quake in the Newark Basin?

If there was even a moderate to major quake under New York City, in the Newark Basin, or even one close by—such as in upstate New York—there would be reason for concern. Much of the island of Manhattan sits on a deep layer of soft soil rich in clay and silt and artificial fills—much of it pumped-out dredged material to enlarge the harbors—over rock-hard bedrock (it's much larger than when it was bought form the Indians by Peter Stuyvesant).

In the case of even a moderate quake, these looser soils and fills would actually amplify the shaking, especially during a quake greater than magnitude 6.0. Such a shaking would cause liquefaction, the sands and silts losing their frictional resistance and turning into a "soupy" surface. This would be especially true for those living west of Broadway between 57th and 96th streets, an area where the soft soils have about a 50 percent chance of turning liquid during a quake. In addition to that location, you don't want to be in the eastern part of the Bronx, Flushing, or obviously, Rockaway, site of the two largest quakes so far.

Next, add in the fact that the city has one of the highest concentrations of high-rise buildings per square foot than anywhere else in the country, with most built without the seismic building code guidelines that came into effect in February 1996. The safest building during a quake will be a steel-framed high-rise; another "best," of course, would be a building already designed with earthquakes in mind. Second best would be the classic prewar apartment houses. They are generally sound, their fronts built with brick and the skeletons made of welded steel. But there are some caveats: ornamental structures such as gargoyles, parapets, arches, and air conditioners would be fodder for the quake, smashing on the streets and sidewalks below. Even such items as water tanks on roofs, even in a moderate quake, would crack or tumble to the ground.

At the other end of the spectrum, brownstones are almost guaranteed to turn to rubble in a moderate quake; one scientist stated that brownstones are comparable to Armenian mud huts in a strong quake. The problems lie with construction—most of New York City's unreinforced-masonry brownstones were built by hand over 100 years ago, making them vulnerable from the top of the roof to the foundation. As one newspaper recently wrote, "When the next quake hits, modern high-rise tenants should sit tight, while older-apartment-house dwellers should stay off the sidewalks, and brownstone residents should run for their lives."

In the end, no matter the building, the damage and problems

would soon catch up to everyone after a moderate or major quake. Pipes, sewers, and gas lines in the streets below would rupture, producing fires and explosions. In fact, the potential for fire increases with every degree of magnitude. Hundreds of water pipes would burst, releasing water into streets and lowering the pressure needed to squelch fires in the higher floors of burning buildings. Though there are redundancies in the water system and there is an extensive water supply from New York's two water tunnels—they carry the water from upstate reservoirs into the city—if either or both are affected by a quake, all bets are off. In fact, one water tunnel comes close to the Ramapo fault in Wappingers Falls, an area that experienced a quake once—a magnitude 3.3 on June 7, 1974. One saving grace may be the city's fireboats, which could pump water from the East and Hudson rivers along hoses that can stretch for miles. But could they take care of so many problems all over the city?

The fire stations in New York City are also a problem: many are housed in unreinforced-masonry buildings. If they could get out, emergency search-and-rescue crews would be strained to keep up with the calls. That is, if you could call—telephone lines and cellular phone towers would probably have their own set of failures, not to mention the mega-increase in calls that would jam the local cells, all unable to handle the volume of calls. Business would come to a standstill, as would transportation systems—in and out of the city. It would be a catastrophe beyond the scale of even New Orleans's horrific 2005 flooding from Hurricane Katrina.

Who's Listening?

Whether in New York City or upstate New York, not many people are ready for a major quake. I have to admit—my home in upstate New York wouldn't withstand such a beating. The already minute cracks in the cement wall outside would crumble; a pinhole in the chimney would open wide; and the creek out back might even change, victim to slides along its banks. Though living in the country does have its advantages: we're already prepared for most major disasters around here—floods, hurricanes, and blizzards. We could probably survive a quake as long as it was less than Moment Magnitude M_W 6.0. At least we hope so.

If you want to talk psychology, numbers are the reason why most people in New York ignore the threat of a moderate to major quake in the state. With only one magnitude 6.0 and two magnitude 5.0 quakes in the past, no one is very worried. In California, quakes of that size barely make the news—they are common in the West. The translation here is that such quakes would be just a drop in the proverbial hazard bucket. People are more concerned that in a century, the warming of global surface temperatures will raise sea levels, causing water to inundate the city more readily during flooding.

But statistics have to rear their ugly heads. Even conservatively, it is estimated that a Moment Magnitude M_W 5.0 or larger quake is a 500-year recurring event in New York; a Moment Magnitude M_W 6.0 is about a 600-year event; and a Moment Magnitude M_W 7.0 or above is about a 2,000-year event. It's true—statistics are based on data gathered, and there isn't much data about New York quakes. And similar to flood-recurrence intervals, this does not mean that a Moment Magnitude M_W 5.0 quake will happen once every 500 years. It means that a specific magnitude quake has a 1 in 500 chance of being exceeded in any given year. These statistical numbers are only meant to be a guide to future possibilities, and therefore, such a quake can happen next year, in 500 years, or anytime in between.

Earthquake Emergency Services

What could happen to people living in New York City after a moderate to major quake? Scattered debris in the streets would make the logistics of getting people from one place to another almost unfathomably difficult. None of this takes into account other factors that can slow rescue and safety efforts after a moderate or major quake. There are also the particularly vulnerable people—from small children, the elderly, disabled individuals, and non-English-speakers, to those who are incarcerated.

Hospitals would reach the saturation point in a short time after the quake, the first wave of injuries resulting from things such as concrete or glass falling on people, causing fractures and contusions to all parts of the body. The next wave would be those needing help for burns from fires and respiratory problems from fumes, and even people affected by the stress of the shaking, such as those suffering from myocardial infarctions, or heart attacks.

Not only is the population high in the city, but obviously, so is the number of buildings. This leads to the next phase of hospital care after the quake: the "crush component," as it is called in earthquake medicine—when a building or heavy object entombs a victim, cutting off circulation to a body extremity. In such cases, waste particles in the muscles build up, tissue is destroyed, and the kidneys are strained. For those who are caught for longer periods of time, dialysis may become necessary to flush the kidneys and body of built-up toxins—but emergency vehicles trying to get to a hospital could lose precious time dodging debris.

Perhaps one saving grace of the September 11, 2001, attack on the World Trade Center was that it will help emergency personnel cope with earthquakes. In a way, what happened after the terrorist act was similar to what could happen in a quake—but the quake would be on a much grander scale. And right now, city officials are paying attention to what occurred in New Orleans in the aftermath of devastating Hurricane Katrina in 2005. All these lessons—and the new building codes put into effect to design earthquake-proof structures—can only help the city cope with a future quake.

NEW ENGLAND QUAKES
What Makes Plymouth Rock?

We all know the story of Plymouth, Massachusetts: a religious group that wanted to separate from the Church of England and practice their own religion contracted with a group of English merchants to pay for their passage and supplies to a new colony. Thus came the first separatists and their sympathizers to Plymouth aboard the *Mayflower* (1620), the *Fortune* (1621), the *Anne,* and the *Little James* (both 1623). The passengers of these ships called themselves the "first comers"; it was only later that Plymouth Colony's governor, William Bradford, would give them the label we know them by now: "They knew they were Pilgrims."

But what most people don't know is that along with heading for a new life, these travelers to the New World encountered something England rarely had: damaging earthquakes. These intraplate quakes-in-the-wrong-places began to shake the colonists shortly after they landed. One of the worst occurred on June 1, 1638, when a damaging earthquake rocked Plymouth Colony, lasting, according to some accounts, for minutes, with lower-magnitude aftershocks occurring for twenty days after. (It's interesting to note that this "new world" was filled with natural hazards: a plague of insects hit the colony in May 1633 and an epidemic killed 20 colonists that summer; in the late summer of 1635, the first hurricane struck the area; and in the same year as the quake, the severest storm yet struck the colony.)

This first recorded shaking in what we now call North America measured between magnitudes 6.5 and 7.0, and was felt with equal intensity by French colonists on the Saint Lawrence River and the English along the Massachusetts coast. Roger Williams, the founder of Rhode Island, wrote in a letter that this quake was the fifth major one felt by the natives (probably the Narragansetts) in 80 years, called *nanamkipoda,* or "when the earth shakes." Based on historical and physical evidence, the quake's epicenter was likely central New Hampshire—far from a plate boundary. This wouldn't be the only quake the people of Plymouth would have to endure: in 1658, only twenty years later, another major quake would shake the colony; there would be another in 1663. By then, more houses and structures had been built—only to fall prey to the shaking by the quakes.

Quakes continued in New England, mainly around New Hampshire and Massachusetts, or off both states' coastlines. One around magnitude 6.0 was on October 29, 1727, and caused damage from Boston, Massachusetts, to Portland, Maine, with around 24 aftershocks. Another struck on November 18, 1755—another magnitude 6.0 quake called the Cape Ann earthquake. The shaking caused widespread damage inland and along the coast of New England. One crew of a sailing vessel 124 miles (200 kilometers) offshore thought they'd struck a rock right after the quake occurred; dropping the anchor, they found plenty of water, not solid ground. And in Boston, chimneys fell and bric-a-brac on houses tumbled into the streets, especially in the harbor region—an area built on loose sediments and rocks.

Quaker States

Just how topsy-turvy is the shaking in New England? According to the Northeast States Emergency Consortium, an organization that deals with helping communities cope with disasters (natural and otherwise), from 1538 to 1989, New England had a recorded total of 1,215 earthquakes; the total for the entire Northeast was close to 4,500. In the twentieth century alone, more than 200 quakes occurred in the New England area, most ranging from moderate (around magnitude 4.5) to micro. The majority are located from the Lakes Region south along the Merrimack River Valley.

But there have been moderate damaging quakes: two of the strongest to date occurred in New Hampshire on December 20 and 24, 1940—both a magnitude 5.5 and felt throughout a 400,000-square-mile (1,035,995-square-kilometer) area. Near the epicenter, chimneys fell or cracked, as did several foundations; in Concord, 50 miles (80 kilometers) from the quake, offices, schools, and houses sustained cracks and structural damage; and in distant Albany, New York, falling Christmas trees were reported. Other, more recent, New England quakes of magnitude 4.2 or more include the June 15, 1973, quake near the New Hampshire–Quebec border; and the January 19, 1982, magnitude 4.5 quake that occurred west of the highly visited tourist town of Laconia, New Hampshire.

Why the concern about a quake in New England versus someplace like California? Like other states east of the Mississippi River, the loose sediment or hard bedrock under this part of the continent carries seismic waves much farther than in the West. Not only that, most of the intraplate quakes are shallow—which means they won't dissipate before they reach the surface. Scientists know this translates into more damage: a moderate Northeast quake could impact 4 to 40 times the area of a quake in California.

The actual physical attributes of New England don't help, either: this region has a population density 10 times greater than California, with most of the larger cities having a higher population density than comparable West Coast cities. Finally, the Northeast has some of the oldest towns in the United States, complete with nonseismically designed structures, homes, and

The tidewater streams of Charleston, South Carolina, have always been ripe for an earthquake—such unconsolidated sediments would only amplify a major quake. *Courtesy of NOAA.*

infrastructure. A person living or working in an unreinforced-masonry building located on a landfill, harbor sediments, or unstable soil wouldn't have much of a chance in a magnitude 5.0 or higher quake.

A WORLD OF SHAKING
Canadian Quakes out of Nowhere
Canada is nowhere near a tectonic plate boundary. To the west, there is the slipping Pacific plate; to the east, you have to travel to Iceland to find the next plate border, the spreading Mid-Atlantic Ridge. But that doesn't mean central and eastern Canada have missed the impact of an earthquake. The earliest referred to came from native Indian accounts around the mid-1500s; the first

recorded quake occurred in 1663, causing landslides in the Saint Lawrence-Saguenay region; the shaking was felt into what is now New England in the United States.

One of the worst quakes not associated (or even associated) with a tectonic plate border was the Grand Banks earthquake, which shook east-central Canada on November 18, 1929. The magnitude 7.2 quake occurred at 5:02 P.M. local time, at a depth of a mere 12 miles (20 kilometers), a shallow quake with an epicenter about 155 miles (250 kilometers) south of Newfoundland, along the southern edge of the Grand Banks. Though the quake was major, it did little damage in the sparsely populated area—mostly Cape Breton Island experienced broken chimneys, cracked buildings, and landslides blocking major highways. A few aftershocks, one as large as magnitude 6.0, reached from Newfoundland to Nova Scotia, but there was little damage. The big shock from the Grand Banks earthquake became the aftermath: a huge tsunami hit the eastern seaboard as far south as South Carolina and all the way across the Atlantic Ocean to Portugal.

There have been others, too, albeit not as strong: on March 1, 1925, a quake was felt over a large expanse of eastern Canada and, again, the New England states. There were no casualties or injuries, but there was considerable damage in a narrow belt covering both sides of the Saint Lawrence River and extending from Trois-Rivières to Shawinigan Falls. Other eastern Canadian quakes included the 1935 Timiskaming in western Quebec and the 1944 Cornwall-Massena earthquake on the upper Saint Lawrence River—both of which caused extensive damage in the areas surrounding the quakes' epicenters.

Responsibility for keeping track of earthquakes in Canada rests with the Division of Seismology and Geothermal Studies, Earth

Physics Branch, Department of Energy, Mines, and Resources (EMR), Canada. The organization has a network of more than 50 seismographic stations, each measuring the epicenter, timing, and

When Cable Breaks, the Internet Falls

On November 18, 1929, the Maritime Provinces of Canada were hit by an earthquake originating off the edge of the Grand Banks, Nova Scotia. The earthquake not only shook the region, but also caused a then-unusual occurrence: during and after the quake, 13 transatlantic telegraph cables set in the ocean south of Nova Scotia broke—10 that parted in two places and 3 that broke in three places. The breaks occurred all along the steep continental slope that lies just beyond the Grand Banks; none were broken on the continental shelf.

The break records resulted in a tidy, unintentional scientific experiment. Because the cables were attached to devices that collected the transmissions, the timing of each break was recorded by measurements of electric resistance in the cables. Eight of the cables high on the continental slope broke instantly during the quake; the last five snapped in progressive order down the slope, each successive cable taking longer to break than the last.

The interpretation of the breaks took a while, with the first reports blaming the snapping on the earthquake-induced tsunami. But because the last five breaks occurred progressively more slowly until the last break, it could not have been the tsunami's fault—the massive wave would have caused the breaks to occur one after the other in rapid succession. After much discussion—and better technology to understand the continental shelves—scientists realized that massive amounts of sediment mixed with water were probably to blame. These so-called turbidity currents are now known to rapidly slide down slopes, helping to carve submarine canyons, deposit sediment on the flatter seafloor below the continental shelf—and break the occasional cable. (More recently, a researcher proposed that the magnitude of the country's quakes (they also measure quakes in other parts of the world). They produce an annual catalogue of all Canadian earthquakes—from micro to major.

> last cables broke as instantaneous slumping and sliding occurred along the slope, but this has not been proven.) And overall, the huge earthquake, and with a nudge from the tsunami, probably precipitated the turbidity current movement.
>
> But skip all those small transatlantic cables and their impact—modern humans have a bigger problem: on December 28, 2006, an earthquake shook for only a few seconds off the coast of Taiwan, but damaged two undersea data-transmission fiber-optic cables, causing widespread Internet slowdowns or stops and phone outages (which led to a halt in financial transactions in affected areas) in Taiwan, Hong Kong, Japan, China, Singapore, and South Korea. The event caused a mild panic, with many analysts speculating that the lack of access to the Internet would force people in the affected countries to remember what life was like before computers made the Earth a global community. And apparently, few techno-savvy people liked the taste of that former life.
>
> It's not only quakes that can rupture the lines to phones and the Internet. Other culprits include volcanoes, fishing trawlers, ship anchors, and even your hungry, nibbling sharks. For this reason, fiber links are generally built as loops. For instance, FLAG Telecom's North Asia loop runs undersea from Hong Kong to Taiwan to Korea to Japan, then takes another route back to Hong Kong. If one link in the loop breaks, data will automatically be switched to flow the other way around the ring, and customers should not even notice a change. But outages can occur when there are too many broken links in the chain in a small area; where this earthquake shook, there apparently are at least a dozen cables in such an array. And of course, the Far East is not the only area subjected to earthquake-related cable breaks. In 2003, a magnitude 6.8 earthquake in Algeria damaged cables in the Mediterranean, cutting links to France and slowing down Internet access across the Middle East.

Quaking Abroad

The United States and Canada, of course, are not the only countries destined to have quakes in the wrong places. Northern Europe would typically be classified as a relatively stable part of the European continent—an area in which the thought of quakes barely enters anyone's head. Compared to places such as the quake-active Mediterranean (mainly Italy and Greece) or the Turkish and Armenian borders to the south, northern Europe is quiescent. Until recently, no one would have suspected that the seemingly insignificant faults there could cause any problems. Earthquakes seemed rare—one of the largest occurred long ago, in 1356, north of the Alps, and destroyed much of the Swiss city of Basel.

But that all changed on April 3, 1992, when a magnitude 5.3 earthquake struck the city of Roermond, in the Netherlands; it caused extensive damage in the region. It was one of the major historical seismic events in that part of Europe—and gave seismologists something to think about. Yes, quakes can happen in all the wrong places in Europe, too.

The northern part of the United Kingdom isn't exempt from quakes, even though it, too, is far from a tectonic edge. Southeast England is rarely thought of as a seismic zone, but earthquakes do occasionally shake the region—with some large enough to cause significant damage. There have been few in recent centuries, but two of the largest British quakes occurred in the Strait of Dover in 1382 and 1580, along a fault zone that extends beneath the English Channel and into Belgium.

What creates these intraplate quakes in the northern parts of Europe? Though tectonic plate movement has been cited by many, other scientists believe it has more to do with ice. Similar to Canada and the northern United States, the legacy from the ice age may be to blame—the crust beneath much of the northern United Kingdom, Europe, and northern North America springing back and cracking after the ice retreated. In the United Kingdom, it may be from the melting of the Scottish ice sheets; in northern Europe, especially around Finland, it may be the ice sheets that once covered Scandinavia.

Whatever the reason, to help map the movement, the European Space Agency has been collecting radar satellite data since 1992, providing information about ground movements often measured in inches. In 2001, the measurements became even more precise when a new processing technique allowed scientists to determine movements of less than a millimeter per year. This allows mapping over a wide area, to improve risk assessment—not only for quake-prone areas, but for those places with infrequent ground movements.

Shaking Down Under
Australia does not necessarily bring the word *earthquake* to mind—after all, the country sits almost right in the middle of the Australian tectonic plate. But the island continent does shake—and far from any plate boundary. For example, on October 21, 2006, in New South Wales, a quake measuring around magnitude 4.0 struck. This moderate magnitude earthquake occurred about 11 miles (18 kilometers) northeast of Reids Flat, east of Wyangala Reservoir, and 86 miles (139 kilometers) north of Canberra. It was reportedly felt in Orange to the north and in Bowral to the east. Scientists know that an earthquake of this magnitude could cause minor damage within a few miles of the quake's epicenter—in this case, there was little damage, thanks to it being a sparsely populated rural area.

Why does Australia shake? The continent is on the Indian-Australian plate, a crustal chunk that is being pushed north, causing it to collide with the Eurasian, Philippine, and Pacific plates. In fact, to its south, Australia is slowly being squeezed sideways by a fraction of an inch (0.01 centimeter) per year. This collisional stress builds up in rocks over the years until they rupture (most likely along preexisting faults), creating these intraplate earthquakes.

And where does Australia shake? Overall, Adelaide is at the highest risk for earthquake hazards of any capital city, with more medium-size earthquakes in the past 50 years than any other. Eastern Australia doesn't experience as many quakes as the west. The old, cold, hard rocks of western and central Australia do not absorb seismic energy well, so earthquakes in these areas are felt over longer distances than in eastern Australia.

Looking back over the country's short historical record (Australia was populated mostly by native cultures until the early 1800s), earthquakes greater than magnitude 4.0 are fairly common in Western Australia. Though it is only speculation because of the lack of data (and a sparse population), it is thought that a quake in 1906 off the west coast of Australia shook the region with an estimated magnitude of 7.5—which may have been the largest-known quake so far to occur in the region. Australia's largest verified and recorded earthquake—estimated at magnitude 7.2—also happened in the west, in April 1941 at Meeberrie, about 62 miles (100 kilometers) east of Perth.

Though not Australia's largest, one of the more famous quakes, measuring around magnitude 6.9, occurred at Meckering, Western Australia, in 1968. The shaking not only caused extensive damage and surface faulting up to 10 feet (3 meters) high, and was felt over most of southern Western Australia, it literally destroyed the small wheat-belt town (miraculously, there were no deaths). Even today, the shaking hasn't stopped, with a quake greater than magnitude 4.0 occurring approximately every five years in the region. Another good shake was the Cadoux earthquake of 1979, measuring magnitude 6.1—and though not as large as the Meckering event, it still caused major damage to the small town of Cadoux, along with a surface rupture about 9 miles (15 kilometers) long.

Australia may not seem to be much of a hotbed of quake activity, but the misplaced quakes can cause extensive damage and loss of life. Yes, most scientists believe that a quake larger than magnitude 8.0—most often associated with subduction zones—is unlikely ever to occur within Australia's borders. And yes, active plate boundary areas like Japan, the Philippines, or California usually experience earthquakes of magnitude 7.0 every few years. But it is estimated that an earthquake exceeding a destructive magnitude 7.0 occurs somewhere in Australia every 100 years or so—which means Australia is not completely out of the proverbial earthquake woods.

More Southern Hemisphere Quaking

Africa's north to central eastern coast has been a bevy of earthquake activity in the past—and present. One such quake occurred on December 5, 2005, when a fault along the East African Great Rift Valley ruptured below the surface of Lake Tanganyika. The earthquake, estimated at between magnitude 6.3 and 6.8, caused widespread damage and was felt in Burundi, Tanzania, Rwanda, and as far away as the coastal city of Mombasa, Kenya. This quake (and others around the Great Rift

Valley) seems easy to understand—the region is right where the land is pulling apart, perhaps trying to create another ocean. It is near plate boundaries—a place where quakes are common.

But it was not the same for the February 23, 2006, quake that measured an estimated 7.5 in southern Africa. At least four people died and dozens were injured as an earthquake shook the area in central Mozambique around midnight local time—a quake that turned out to be the most powerful in the area for more than a century. The earthquake was felt by people throughout Mozambique, as well as in parts of the neighboring countries of Swaziland, Zambia, Zimbabwe, and South Africa. Power outages were reported, and though most of the larger cities escaped serious damage, traditional houses did not fare as well, collapsing in such towns as Espungabero.

Though it was south of the Great Rift Valley, this major earthquake was unusual and unexpected in southern Africa. Seismologists expect events in the magnitude 7.0 range to occur about every 50 years in this area. In fact, an earthquake similar in size and in a similar location occurred in 1940. Such quakes are felt for long distances—up to 620 miles (1,000 kilometers) from the epicenter—because of the region's unique rocks: the mostly homogenous rock above the junction of the South Africa plate and Somali plate has few obstacles that would stop traveling seismic waves.

And then there was the December 22, 1963, magnitude 6.4 quake that hit the other side of the continent—yet another quake in the wrong place. This quake had its epicenter northwest of Guinea—an intraplate locale with a historically low incidence of quakes. Scientists believe that this unusual quake occurred on a preexisting fault—but don't know how the shifting started. The results included a surface rupture, extensive rockfalls, and even minor liquefaction effects at distances less than 6 miles (10 kilometers) from the surface faulting and main shock epicenter.

What about the other big Southern Hemisphere continent, South America? The west coast of the teardrop landmass has had its share of quakes—and large ones, too. The countries of Peru, Chile, Ecuador, and Colombia lie at the site of a long subduction zone, in which the Nazca and Antarctic tectonic plates dive under the South American plate. The major result is the Andes mountain range, often called the backbone of the continent, and home to a part of the "Ring of Fire" that surrounds the Pacific Ocean. Thanks to this turbulent region, the largest quake ever recorded occurred on May 22, 1960—the great Chilean earthquake, recorded at magnitude 9.5 (though that number is still considered by some scientists to be too high, by others to be too low), ruptured its fault for 621 miles (1,000 kilometers).

There's another side to South America—the east coast. For the most part, the eastern part of the continent is quiescent compared to the est. Logically, many of the eastern quakes center on the region around Caracas, Venezuela, right near an interplate boundary of the Caribbean plate. For example, a major quake struck the region around Caracas in 1674 (today's Catedral de Caracas stands on the ruins of a former sixteenth-century cathedral that was destroyed by the earthquake). On March 26, 1812, it's estimated that a magnitude 9.6 (though this is highly debated) quake hit Caracas, damaging 90 percent of the city, with fatalities between 12,000 and 20,000—one of the worst natural disasters in the country's history. Thanks to the then-turbulent history of the region, the Spanish clergy in Caracas claimed that the earthquake was God's anger against the sins of the rebel government. (It's interesting to note: some of the worst earthquakes ever to hit North America—the New Madrid quakes, which ran from December 1811 to February 1812—ended just the month before. Was a tectonic plate going through a megashifting event at that time? No one really knows—and no one is saying that the two quakes were connected in any way.) Yet another quake occurred at midnight on July 30, 1967, again in Caracas, measuring magnitude 6.5, killing 277, and causing millions of dollars in damage. One of the more recent big quakes occurred on July 9, 1997, measuring a Moment Magnitude M_W 6.9. The quake mostly affected the towns of Cumaná and Cariaco, Venezuela, and left more than 80 people dead, nearly 400 injured, and 600 homes destroyed.

But Brazil's small tip of land that juts into the Atlantic Ocean and parts of the Amazon Basin seems to have numerous quakes—all of which occur at an intraplate region. For example, an earthquake measuring Moment Magnitude M_W 7.0 occurred late at night on June 19, 2003, in the Amazon jungles of Brazil. According to the Seismological Observatory of the University of Brasília, the quake took place along the Rio Araguaia, just south of the Serra do Roncador mountains, about 331 miles (553 kilometers) west of Brasilia. Because it was a sparsely populated area of the jungle, few people felt the shaking; in addition, the quake focus was deep—around 345 miles (555 kilometers) below the surface. But seismologists took notice—Brazil had not experienced an earthquake of this magnitude for 20 years.

Not too far up the road, a magnitude 5.2 quake struck French Guiana, 30 miles (50 kilometers) southeast of the capital of Cayenne, at around 1:30 P.M. local time. Though there were no reports of major damage or injuries, some buildings were evacuated, people poured into the streets (some fainted), and the fire department received hundreds of calls. Some slight tremors were also felt in Paramaribo, the capital of neighboring Suriname, about 240 miles (385 kilometers) east of where the quake hit. The part that made scientists take notice of this quake in the wrong place: the USGS noted that a quake hadn't been detected within 185 miles (300 kilometers) of the area in at least 33 years—and some reports stated it was the first time a quake had been felt in Suriname in recent memory.

What are the chances of a major quake occurring in this region of the country? According to seismologists, there is always a chance that a moderate to large quake could hit, even in this intraplate region. Because of the difficulty in accessing the Amazon and nearby regions, little is known about existing faults with the potential to cause quakes. Add to this the faults on the Atlantic Ocean floor, which could move and create not only a quake, but a tsunami. There's a great deal at stake, especially for the eastern coastal towns—including Rio de Janeiro to the southeast.

Shaking the South Pole
The continent of Antarctica—the size of the United States and Mexico combined—may be considered one of the highest, driest, and coldest spots on the planet, but it's also a landmass that only *seems* to be devoid of earthquakes. Most bona fide quakes occur in Antarctica's active western peninsula, where they strike every two to three days (usually microquakes); whereas temblors are rare in East Antarctica. During the last century, only three were known to have hit the region—and those were tele-seismically recorded using instruments on other continents. No one truly knows what

causes the quakes yet—but seismology on "the ice" is a very "hot" study these days.

Just how Antarctic quakes occur has long been a matter of conjecture—especially those recorded in the center of the aptly named Antarctic plate. Some smaller tremors have been attributed to the calving of ice sheets or the fracturing of the ice. Other temblors also occur around Mount Erebus, and other young volcanoes along the western Ross Sea—evidence that tectonic activity is presently occurring.

What about the quakes recorded in the Transantarctic Mountains along the western Ross Sea? This unusual intraplate mountain belt, wholly within the Antarctic plate, seems to be rising each year. Are they the result of uplift along the edge of a rift system that developed within the plate during the fragmentation of the Gondwanaland supercontinent (the southern part of Pangaea) millions of years ago? Or do they lie over a mantle plume or hotspot similar to what is forming the volcanoes of Hawaii and Iceland? (Scientists believe there may be a hotspot under the eastern margin of Marie Byrd Land that also resulted in a sub-ice volcano far to the south, but most scientists don't believe such a Transantarctic plume exists, as it would have to be unbelievably large to cause the uplift.)

And there is Lake Vostok in East Antarctica: this frozen lake may not be sitting on a quiet craton—a tectonically stable part of a continent—as most scientists have thought. Airborne reconnaissance and seismic studies of the icy continent recently showed that the land under the world's largest subglacial lake is an active geologic boundary—and studies are currently under way to prove it.

That doesn't mean major quakes don't hit the huge icy continent. For example, a large Moment Magnitude M_w 8.1 earthquake occurred off the southeast coast of Antarctica near the Balleny Island region on March 25, 1998. The quake occurred in the midplate region of the Antarctic plate. But the weirdest part was the lack of reports of such large earthquakes in this region. Scientists know that the mechanism that caused the massive quake could not be a tectonic event. In fact, some researchers believe that the quake may be the Earth's response to present-day and past ice mass changes in Antarctica. After all, ice is just another type of "rock."

Recording Quakes at the Southern Polar Region

Pity the poor seismologists who work in Antarctica—a land where gathering earthquake data is a major task. Why is it so difficult to detect and study earthquakes in Antarctica? Because the operation of seismic equipment—and where those devices have to be anchored—and the actual collection of data are difficult. Each seismic instrument must be able to withstand the extreme climate of the ice sheet in temperatures that fluctuate from just above freezing to close to a hundred degrees below zero.

In addition, power for the instruments becomes an issue: I've been in Antarctica during the Southern Hemisphere's summer, and for three months, the Sun truly shines 24 hours a day (you can't pull "all-nighters" here). This allows the seismic stations to operate on solar power during the summer; but when the winter months arrive, wind generators must charge the batteries that supply the power to heat the stations and operate the equipment—but only *when* the winds blow.

Satellites are also a problem. In most other parts of the world, seismic data can be sent via satellite from distant locales, but in Antarctica researchers must return to the seismic stations to download the information. This is because satellites are not in orbits that allow the information to be readily relayed to researchers at a distance. In fact, at the South Pole—and even McMurdo Station—we experienced an Internet "blackout" each day as the satellites moved out of range.

CHAPTER 8

Bad Effects

> *Strange rumbling sounds were heard here and at Tabradden and elsewhere on Monday morning early, and a most unusual rise of tide took place on Monday evening at eight o'clock, and again at half eight [sic]. The sea receded over 100 yards, enabling people to pick up fish in quantities. The draw back was of short duration. The sea returned with a great rushing noise and those who were fish gathering had to run quickly for their lives. Boats that were high and dry on the beach were washed out and swamped. The Rob Roy and the lighters at anchor were twisted right around. The sea rose about eight feet. In a short time all was quiet again, except that the ocean continued rising and falling until next morning.*
> —*Extracted from* The West Australian, *Friday, August 31, 1883, around the time a massive volcanic eruption occurred at Krakatau Islands in the Sunda Strait region of Indonesia*

It isn't enough that earthquakes strike along plate boundaries and sundry intraplate regions. There are also the aftermaths—the out-of-nowhere side effects of quakes that destroy and kill people.

The ground rattles around volcanic regions—most of them along plate boundaries, but other volcanic regions can violently shake, far from such borders. Walls of water in oceans and lakes called tsunamis sweep away hillsides and coastal towns as a result of huge quakes—many of which occur hundreds, if not thousands, of miles away from the initial quake. Shake again, and clumps of land slide down mountainsides, road cuts, and hills, burying towns, highways, and people, even after a minor quake. These are the bad effects—disastrous events that can occur along with a quake, even those in the wrong places.

VOLCANIC SHAKES
Volcanic Ways and Means
We've all seen the B movies: a geologist as hero/heroine arrives at a secluded spot in the mountains or along a beach. A person at the local barbershop/grocery store tells the scientist that there are strange things happening around the town, things that have never happened before. The geologist just happens to find some geologic evidence—usually nonscientific, but terrific in terms of Hollywood—that leads him/her to believe an eruption is imminent. The ground shakes, the buildings tremble and fall, and a nearby mountain (or beach) suddenly explodes. Some residents will make it—no doubt the hero and his girl or heroine and her man—but the bad guys will be wiped out by a lava flow. And there's *always* a lava flow.

Switch to real life Mexico in 1943: for three weeks in late winter, people near the village of Paricutín heard rumbling noses and felt a multitude of earthquakes. On February 20, Dionisio Pulido, a Tarascan Indian farmer, decided to collect shrubbery, readying his field for spring sowing—when the ground split open for about 150 feet (46 meters), with a depth of about a foot and a half (half a meter). In his own words, he then "felt a thunder, the trees trembled, and I turned to speak to Paula [his wife]; and it was then I saw how, in the hold, the ground swelled and raised itself two or two and a half meters high, and a kind of smoke or fine dust—gray, like ashes—began to rise up in a portion of the crack that I had not previously seen. . . . Immediately more smoke began to rise with a hiss or whistle, loud and continuous; and there was a smell of sulfur. Then I became greatly frightened and I tried to help unyoke one of the ox teams. I was so stunned I hardly knew what to do . . . or what to think."

This was the beginning of Paricutín, a cinder volcano that in its first day grew 165 feet (50 meters); within a week, it rose to a 330-foot- (100-meter-) high cone. Its activity was brief in terms of volcanoes, with the last eruption ending in 1952. By then, it was a 1,391-foot (424-meter) volcanic mountain, its extensive lava fields covering the towns of Paricutín and San Juan Parangaricutiro—in all, the ash and lava covering 0.3 cubic miles (1.4 cubic kilometers). Amazingly, only three people were killed during the eight years of eruptions—all from pyroclastic flows from the volcano. But the ash and lava still had a long-term dramatic effect, smothering agricultural lands and changing a way of life.

It's true that Paricutín sits along a swath of over 1,400 volcanic vents—the mostly inactive (or, scientifically, a monogenetic field) Trans-Mexican volcanic belt. But there was never an indication that such a mountain would shake and rock into existence—only two volcanoes have ever grown right in front of human eyes in North America during historical times: Paricutín and, in 1759, Jorullo, a volcano that appeared in western Mexico, about 50 miles (80 kilometers) southeast of Pulido's farm.

Volcanic eruptions certainly have the ability to move the Earth—testimony to the molten rock squeezing up from below and through weak spots in the Earth's crust. Many volcanoes are associated with major earthquakes and tremors, the majority found around plate boundaries—and most of those in association with subduction zones. The past two centuries have been witness to at least a dozen volcanic eruptions in these zones, mostly in smaller, third-world countries, such as Mexico, Indonesia, Colombia, Martinique, Papua New Guinea, and the Congo, to name a few. The most lethal in the twentieth century was in Martinique in 1902; but it was the nineteenth century that led the volcanic pack—the 1815 Tambora eruption affected the global climate, and in 1883, Krakatau in Indonesia blew apart an entire island and sent volcanic ash drifting around the world for years.

Why the concern? The volcanoes that wake up after lying dormant for centuries and those that appear out of nowhere cause the most consternation. If the eruption that tore apart Tambora were to occur today—near or far from a plate boundary—the devastation would be tremendous. The population has increased and changed, and tens of thousands of people could die, instead of only hundreds in a sparsely populated region. And it's not only the ash and lava that concern scientists—it would also be the tremendous earthquakes that often accompany major eruptions.

Volcanic Details

Around 10 percent of the world's population—about 600 million people—live in the vicinity of one of the Earth's approximately 1,500 active volcanoes. Not all of these volcanoes are the same. For example, *stratovolcanoes* experience violent eruptions; those with lava flows experience less explosive, more oozing eruptions.

But overall, most of these mountains have the potential to spread deadly ash and/or lava.

The majority of volcanic mountain chains on land and island arcs in the oceans form mainly due to subduction zones—as one plate slips under another, the friction from the movement, and heat and pressure from depth, cause the rock to melt. Simply put, this lighter, melted magma rises, following weaker spots in the rock layers, eventually reaching the surface and forming a volcano. For example, Seattle, Washington, lies within the destructive circle of 14,413-foot- (4,393-meter-) high Mount Rainier, a volcano in the Cascade Range, a result of the Juan de Fuca plate slipping under the North American plate. The huge volcano El Popocatépetl is visible on the skyline of Mexico City, Mexico—the world's largest city, with about 20 million people living in the shadow of the volcano. This area is where the Cocos plate dives down under the North American plate—and just offshore is the evidence: the Middle America trench. Alaska's Aleutian Islands, an arc of volcanic islands, formed as the Pacific plate subducted under the northern part of the North American plate. And the list goes on.

Midocean (or submarine) ridges also spawn volcanic islands, the resulting rift volcanoes located where plates are separating. The classic example is Iceland, a large island formed as the result of seafloor spreading in the Atlantic Ocean—the Mid-Atlantic Ridge that separates the European and North American plates. And finally, volcanoes can pop up around areas known as hotspots—right in the middle of a tectonic plate (read on for more about hotspots). The classic example—and one that occurs right in the middle of the Pacific plate—is the Hawaiian Islands chain.

The Other Sides to a Volcano

Dangers from volcanoes are many, including the devastating impact of such activity on nearby landscapes and communities. Deaths are often the result of suffocation (by volcanic ash or gases, as in the famous case of Pompeii, Italy), lava flows, erupting bombs, or even shock waves from the eruption. Certain types of volcanoes cause the most casualties: based on past records and current observations of volcanoes, stratovolcanoes (explosive volcanoes) are the worst culprits. Statistically speaking, there are more stratovolcanoes than any other type, meaning more people live around such mountains.

But there are other indirect effects, too—including the dangerous mudflows called *lahars*. These volcanic-debris flows carry some of the worst effects, dropping poorly sorted material along valleys near the explosive volcanoes. They most often occur when heavy rains wash loose materials from earlier eruptions or when volcanic material is shaken loose during an earthquake. For example, the 1991 Pinatubo eruption in the Philippines was the second largest of the twentieth century (after Katmai in 1912) and deposited a huge amount of loose material on the steep and gullied slopes of the volcano. Add to this the high precipitation in the Philippines, and it is easy to see why hundreds of lahars have been generated since the initial explosion. In fact, since 1991, more people have been killed or injured due to lahars than due to the original eruption.

The list of indirect destruction by volcanoes goes on: thick glaciers or winter snows on volcanoes melt during an eruption, causing massive amounts of water to flood the valleys below. Airplanes can be affected by the huge ash plumes sent up by an erupting volcano; oftentimes, the plumes are at the mercy of local weather conditions. And depending on the amount of volcanic debris sent into the atmosphere, the local—or global—climate can be changed for a short time after an eruption.

VOLCANIC SHAKING

Earthquakes and volcanoes are definitely linked through plate tectonics, with the majority of seismic and volcanic belts resting together between plate boundaries and through volcanic regions outside plate borders. But many of the questions about earthquake–volcanic interactions remain unanswered—the field is in its infancy, and the lack of data slows the progress toward answers.

What scientists *do* know is that all volcanoes have one thing in common: movement. In the majority of cases, before, during, and after a volcano erupts, some type of shaking occurs. This volcanic shaking differs from regular fault-caused quakes caused by a sudden release and rapid decrease in seismic energy. In an active volcano, there is a more or less continuous ground vibration called volcanic tremor—not a shaking felt by most people, but a kind of hum detected by seismographs. Though the precise reason for this shaking is unknown, it is thought to be from irregularly moving magma and its pressure on the surrounding rock; temperature changes as the magma flows through the rock; or the rumbling from forming and exploding gas bubbles—from groundwater, magma, or both—as the liquid rock moves through the paths of least resistance.

Whatever the source, in many volcanoes, the tremors occur during an eruption, usually beginning before the magma reaches the surface. This does not happen with all volcanoes, but with the ones that exhibit a swarm of what are called long-period earthquakes, it is a reliable sign; these precursors could indicate an eruption in days or even weeks. For example, for active volcanoes in Hawaii like Kilauea, high-amplitude microquakes indicate the start of an eruption—or at least one beginning within hours. (At the Hawaiian Volcano Observatory there are even automated "tremor alarms" to warn scientists of any quakes that portend potential eruptions.) With volcanoes associated with subduction zones, such as Mount Saint Helens in Washington, tremors during quiescent times can be about a few per month, whereas swarms before an eruption can increase to several hundred per day. The Redoubt volcano in southern Alaska and the Galeras volcano in Colombia have both recently exhibited multiple microquakes just before an eruption.

There is still disagreement about how much these microquakes can tell volcanologists about an imminent eruption. Many studies believe that only about half of all volcanoes that have quake swarms have eruptions. A recent Russian study examining the Kamchatka volcano states that large increases in the total energy released by volcanic quakes are a more important indicator of eruption than an increase in quake numbers. Even microquakes have been suggested as an indicator. No one truly knows—maybe it's a combination of all these factors.

And there is a "vice versa" effect: large earthquakes are certainly capable of triggering eruptions within a matter of minutes or days at nearby volcanoes, especially those within a certain distance from the fault on which the earthquake occurred. For example, on November 29, 1975, an eruption on Kilauea was preceded by a magnitude 7.5 quake that struck beneath the south flank of the volcano only a half hour before. The May 22, 1960, great Chilean earthquake—one of the largest ever recorded, at magnitude 9.5—was the probable cause of the Cordón Caulle volcano in the central part of the country only two days later. Some places are doubly affected by different quakes: the Long Valley caldera in eastern California showed seismic and volcanic unrest after the magnitude 7.3

Landers earthquake (June 28, 1992) and the magnitude 7.1 Hector Mine quake on October 16, 1999.

The Hot Debate over Hotspots

They may not be as well known as volcanoes and earthquakes, but hotspots definitely have a place on the list of earth movements in the wrong places. The traditional theory holds that a hotspot is a stationary, localized flow of magma pushing through areas of weakness in an overlying continental or oceanic plate. Unlike most volcanoes found along plate boundaries, hotspots occur in the middle of a plate, creating volcanic features far away from where they would "normally" be found. They can be a spot of not only moving magma, but of small, moderate, and even major earthquakes.

The idea of hotspots was first proposed by Canadian geophysicist J. Tuzo Wilson (1908–1993) to explain the presence of volcanoes in mid-plate. (He's also known as the person who discovered transform faults.) In his hotspot theory, which emphasized

---

Water and Volcanoes

What's the often bizarre connection between a major earthquake, volcano, water—and great distances? Take a shield volcano called Mount Wrangell in south-central Alaska: this 14,163-foot (4,317-meter) mountain responded to the surface waves from the December 2004 Sumatra earthquake—the same one that caused massive destruction from tsunami waves in the Indian Ocean, a shaking that registered an estimated Moment Magnitude M_W 9.2. The powerful quake occurred around 7,000 miles (11,263 kilometers) away, but the volcanic mountain still responded, heating up with small internal earthquakes. Scientists believe the hefty seismic waves compressed the volcanic piping system within the mountain, causing the quaking reaction.

The waterfront at Seward, Alaska, looking south before being devastated by an earthquake, underwater landslides, surge waves, and a tsunami after the great quake of March 27, 1964. *Courtesy of USGS.*

oceanic hot spots, Wilson stated that relatively small, long-lasting, hot regions—like the Hawaiian Island complex—needed a source for their heat. The result was the idea of hotspots—regions usually located at the interior of tectonic plates that provided localized sources of high heat energy. These thermal plumes and their molten rock would provide the material for volcanism, as the lighter magma would rise through the crust and erupt on land or the seafloor.

In the oceans, as the eruption continued, the material would build up, growing into what is called a *seamount*, or underwater volcanic mountain. Over thousands of years, the volcanic feature would grow enough to pop above the sea surface, creating a volcanic island. But oceanic plates don't remain stationary, and as the plates move in their long march across the Earth, the plume

The waterfront at Seward, Alaska, looking north after being devastated by the results of the March 27, 1964, earthquake. *Courtesy of USGS.*

would stay put. As the volcano moved along with the plate, it would eventually outpace its thermal plume, the source of its magma. The island stops spewing molten rock and becomes dormant, but the plume is still active, creating its next offspring—first a seamount again and then an island.

Today many geologists agree that there are such regions that have volcanism far from plate boundaries, but not everyone agrees with the theory of hotspots. Some scientists believe it's not necessary to have a mantle plume to explain such volcanic regions; one recent study suggests that hotspots are shallow phenomena, the magma rising from just under the crust, not from deep in the mantle as in Wilson's theory. Still others don't believe that the thermal plume remains fixed in one spot.

Despite the debate, so far scientists have found about a hundred so-called hotspots beneath the ocean and continental crusts. Some of the more famous examples include those that congregate around spreading plate boundaries, such as those near Iceland; others are far from any plate boundary, seemingly popping out of nowhere, often at the middle of a tectonic plate; the Azores, Canary, Society, and Galapagos islands top the list. The Galapagos Island hotspot outranks the others in several categories: it's associated with a hotspot on the Galapagos rise and a spreading center that extends east from a triple junction with the Pacific rise.

Growing Hotspot: Hawaiian Islands?
In October 2006, an earthquake rocked in what seems like a wrong place—the Big Island of Hawaii. The magnitude 6.5 quake caused windows to rattle, televisions to fall off their stands, and highways and sidewalks to crack. The quake triggered at least one landslide over a major highway, knocked out power to thousands across the entire stretch of islands, and cut off most phone service. A state of emergency—along with the National Guard, which was coincidentally conducting maneuvers there—was called. The quake's focus was around 24 miles (39 kilometers) deep, near the town of Kailua-Kona on the west coast of the Big Island, and was thought to be the result of quakes from the island's active volcanoes. Dozens of aftershocks followed, one reaching a decent magnitude 5.8. Determining the damage was difficult for a few days, as a heavy rain was falling in the region. But later, everyone knew—the earthquake had caused injuries, landslides, power outages, and airport delays and closures, and property damage of around $200 million.

One thing was for sure: hefty quakes are a rarity for these islands found in the middle of the Pacific Ocean, far from a plate boundary. Major quakes occur sporadically around the islands—a magnitude 6.7 in 1983, a 6.1 in 1989, and the biggest, a 7.9 in 1868. Interestingly enough, the islands have probably experienced more tsunamis than quakes.

As stated above, the most accepted theory of why the intraplate islands form is still just a theory: the Hawaiian Islands supposedly formed as a result of a long-lived hotspot under the Pacific plate, located deep under the center of the Pacific tectonic plate. Over a period of approximately 75 million years, the plate has moved slowly over the stationary hotspot, resulting in a chain (or "track") of almost 200 volcanoes extending in a northwest direction. Most of the vol-

The Alaska earthquake on March 27, 1964, caused a major debris avalanche on the peninsula between Ugak and Kiliuda bays. A slide of Tertiary rocks from the 1,500-foot-high peak at the upper right flowed into the uninhabited valley below at about 300-foot altitude, where it spread out as a debris lobe roughly 1,500 feet across. The narrow streak of light-colored debris in the lower right corner is part of the slide that overflowed the near flank of the landslide scar. *Courtesy of USGS.*

The horrific result of the Sumatran tsunami in the Indian Ocean, January 2, 2005. A village near the coast of Sumatra lies in ruins after the tsunami struck Southeast Asia. *Courtesy of the United States Navy.*

canoes are now located beneath the ocean's surface; the ones that rise above sea level constitute the present-day Hawaiian Islands. The "farthest"—and thus oldest—island is Kauai, the northwesternmost inhabited. It is about 5.5 million years old. The most recent creation, and one that may still be over a hotspot, is the Big Island of Hawaii, at the southeastern part of the island chain. Its oldest rocks are less than 0.7 million years old, and its youngest are forming even as you read this book. Still farther out off Hawaii is another forming volcanic mountain—currently a growing seamount called Loihi.

Where does this possible hotspot originate? One study recently proposed that the answer may lie along the boundary between the mantle and the metallic core of our planet. The researchers' data showed that a magma plume may be located approximately 1,800 miles (2,900 kilometers) beneath the crust, where rocks at the base of the mantle are heated by the molten outer layer of the core. The scientists also believe that the magma plume first flows horizontally toward the base of the hotspot, then rises vertically toward the surface—perfect conditions to grow volcanic islands.

> ### Sidebar: An Infamous Ancient Hotspot
>
> Millions of years ago, there were probably many more hotspots, and the Earth overall was warmer than today's "cooler" planet. The crust was thinner, with the moving mantle closer to the surface, allowing the hot magma to well up in more places. Scientists know of one such famous hotspot located in the Indian Ocean; what is now India was located above this hotspot approximately 67 million years ago. Around this time, the dinosaurs were in rapid decline, and large amounts of lava (molten rock) erupted, producing an extensive volcanic field called the Deccan Traps.
>
> Over time, the Indian plate moved slowly to the northeast. The hotspot "stayed put," and like a giant conveyor belt belching out magma, the crust rode over the magma, creating the various islands and geologic features we see today: the Maldives formed around 57 million years ago, the Chagos Ridge approximately 48 million years ago, and the Mascarene Plateau some 40 million years ago. Eighteen to 28 million years ago, the Mauritius Islands formed, and most recently, in the last 5 million years, Reunion Island was created, made up of the Piton des Neiges and Piton de la Fournaise volcanoes.

Will the Earth Move at Yellowstone?

Perhaps one of the most controversial "hotspot" spots is on land: Yellowstone National Park. In addition to the debate over whether it is a hotspot, some scientists believe it is a volcanic area with the potential to erupt in the future. The region lies far from a tectonic plate boundary—partly in Idaho and Montana, but mainly in northwestern Wyoming—and is considered a hotspot because of its past and present volcanic history. Could this be a hotspot in the "wrong" place?

The entire area encompasses a volcanic plateau standing about 8,000 feet (2,400 meters) above sea level, all tucked into the Rocky Mountains. Not that the region has been quiet over time. At least three huge volcanic eruptions occurred around Yellowstone within the past 2 million years, the most recent occurring about 600,000 years ago—all combined with at least 1,000 times more power than the Mount Saint Helens eruption in 1980. Lava spread for thousands of square miles as the large underground magma chamber emptied. Ash deposits from the various eruptions spread as far away as Missouri, Texas, Iowa, and northern Mexico. The chamber eventually collapsed, forming a caldera 30 miles (45 kilometers) wide, 45 miles (75 kilometers) long, and several thousands of feet deep. As the crater filled in with lava over time, it created the central Yellowstone basin we see today.

Since that time, the Yellowstone that we all know has kept its hot disposition. More than 10,000 hot pools and springs, geysers, and bubbling mud pots (pools of literally boiling mud) dot the region. Beneath the ancient caldera is still a magma chamber containing hot molten rock; above it lies a hydrothermal system that supplies the area with water to create hot water and pressurized steam—providing the image we usually think of when we hear the word *Yellowstone*.

The ground around Yellowstone also "breathes" or heaves periodically in various places around the park—sometimes as much as a half an inch (one centimeter) per year. Though this may not seem like much, it is an indication that Yellowstone is indeed still an active volcanic region. The biggest debate right now is whether or not this movement—or any changes, such as differences in earthquake

Top of falls at East Lae'apuki ocean entry, Kilauea Volcano in Hawaii. Looking west, with slopes of Pulama pali in the background on February 21, 2005. *Courtesy of USGS.*

tremors, or geyser or mud-pot activity—is an indication that Yellowstone will eventually erupt again. Most scientists at this time believe such changes are merely hydrothermal pressure changes, and not an indication of renewed volcanic activity in the area.

Yellowstone Source?

Today's Yellowstone Park is the end of a chain of many "Yellowstones" in the region—areas now cooled in a line running southwest from Yellowstone all the way to Nevada. These older plumes formed over many millions of years, with the last giant eruption in the chain taking place around 600,000 years ago.

One of the strangest recent discoveries about the modern

Sunset Crater is a dormant volcano located in Arizona. *Courtesy of USGS.*

Yellowstone plume took place after scientists made three-dimensional images of magma under the park. Using data from around 80 seismic stations that monitor the shaking around the park, researchers developed tomographic images of the melted magma. Strangely enough, the current plume does not seem to rise from the core-mantle boundary (about 1,700 miles [2,700 kilometers] deep) as previously thought—it seems to appear out of nowhere between 310 to 400 miles (500 to 650 kilometers) below the surface. The plume travels under the Montana–Idaho border, northwest of the park, and finally tilts to the southeast, rising through the mantle until it's directly under Yellowstone. No one really knows the source, though some scientists now believe the nearby Columbia River flood basalts may be a clue, as it formed about the same time that the Yellowstone plume formed the series of earlier Yellowstones.

During the Loma Prieta, California, earthquake on October 17, 1989, the Watsonville area experienced ground shaking—triggering liquefaction in a subsurface layer of sand. This caused lateral and vertical movement in an overlying layer of unliquefied sand and silt, which moved from right to left toward the Pajaro River. This type of ground failure is termed "lateral spreading" and is a principal reason for liquefaction-related earthquake damage. *Courtesy of USGS.*

> ### Could Yucca Mountain Quake?
>
> The nuclear waste facility at Yucca Mountain in south-central Nevada remains as controversial as it was the first time it was mentioned as a possible candidate for buried waste material. No one really wanted it there; no one wanted it in their own backyard. But the site was chosen for a variety of reasons—mainly its distance from populated areas and its supposed geologic stability. After all, the site would have to exist for hundreds of years; make that thousands when it comes to the nuclear wastes finally not being radioactive. It seemed to be a waste site in the right place.
>
> But no matter how hard everyone tried, there was no way to deny it: there are faults under Yucca Mountain, just like everywhere else in this world. Talk about a possible earthquake in the wrong place.
>
> Why did they nevertheless choose Yucca? Those responsible for the site's location proposed that the area was not geologically active, and that the groundwater was sufficiently underground that even if there was a containment problem, the wastes would remain inside one area. But not everyone agrees. In 1997, a study out of the University of Colorado at Boulder showed that a moderate earthquake in the vicinity of the repository could cause groundwater to surge into the storage area. The scientists based their findings on historic quakes, geologic data, and results from 20 test wells. They showed that a magnitude 5.0 or 6.0 earthquake

TSUNAMIS
Striking Waves

A good friend of mine once told a tale about patiently waiting at a bus stop in Hilo, Hawaii, on April 1, 1946. It was one of those beautiful early mornings Hawaii always seems to produce—sunshine, wind, and surf on the windward side (east coast) of the Big Island. But the bus was late; being the conscientious kind—

could raise the groundwater levels at the storage site between 450 to 750 feet (137 to 230 meters).

They believe that open fractures under the mountain have allowed the groundwater to descend to abnormally low depths; closed-off fractures to the north have created more normal water tables, at about 600 feet (183 meters). They reason that if an earthquake occurred, it could squeeze shut the open fractures, causing the water to fill in and upward into the storage facility. They cite such occurrences in other places, including the magnitude 5.6 earthquake that occurred at Little Skull Mountain near the repository. In 1992, this area reported a major change in groundwater levels, the quake causing the groundwater to squeeze upward. Other places have reported not only a rise in water levels, but even water erupting from the surface.

Probably the most disturbing evidence is geologic: the region near Yucca Mountain is definitely tectonically active, with volcanoes within a few miles of the site. If a quake from this active region occurred and water reached the storage area, it could cause a rapid corrosive breakdown of containers; plutonium could leak into the groundwater, spreading and making the local area unlivable. Such contamination of the groundwater would eventually leak to the surface, the evaporation causing toxic materials to enter into the atmosphere. With the repository set to contain the wastes for 10,000 years, one or more moderate earthquakes could cut that time down considerably.

and a new teacher at a school that believed in promptness—my friend decided to walk up the winding hill to the school and greet her class on time.

Only a few minutes after she reached the school, she heard the news: not a half hour after she left that sidewalk and bus stop, a huge seismic wave—a tsunami—hit the beach, one of at least six arriving in 15- to 20-minute intervals that morning.

The rushing waves, some estimated to be over 50 feet (15 meters) high in parts of Hilo, and lower in others, ran past the bus stop and up several streets, tearing away not only sidewalks and signs, but houses, businesses, and people. To the day she died, my friend credited the school's strict schedule and rulebook for saving her life—not to mention that the school was sited on a hill far above Hilo's beach.

The tsunamis that struck the Hawaiian Islands were generated by an earthquake that occurred about 15.5 miles (25 kilometers) below the crust. It was not a quake around the islands—it occurred nearly 3,000 miles (4,827 kilometers) away, near the Aleutian trench along the Aleutian Islands, Alaska (more precisely, off Unimak Island). The quake was first translated to be magnitude 7.2 on the Richter scale; based on the energy data, scientists later changed the number to the more current Moment Magnitude scale, the quake measuring 8.6 M_W.

During the quake, a large section of the seafloor was uplifted along the fault, displacing the water and creating Pacific-wide tectonic tsunami waves. Like the effect of dropping a stone in a pond, the waves reached out in every direction from the earthquake's epicenter, spreading across the vast Pacific Ocean. The first to feel the impact was the five-man Coast Guard light station at Scotch Cap—obliterated by the foaming 100-foot (30-meter) juggernaut at 2:13 A.M.

Less than five hours after the shaking occurred, the first of the tsunami waves struck the Hawaiian Islands. The waves all came without warning—there was no Pacific Ocean warning system in place yet—and killed over 170 people. The water pushed inland, destroying Hilo's waterfront, with every house on the main street facing Hilo Bay ripped off its foundation, carried across the street, and smashed into the buildings on the other side. It is here, too, where my friend once stood. She would still shiver when she remembered how she visited the spot a day after the tsunami. There was no sidewalk, sign, or bus left where she had stood that fateful morning.

Monster Waves out of Nowhere

Tsunami is a Japanese word roughly translated as "harbor wave"; it is now used internationally to describe waves generated by earthquakes that travel for hundreds of miles across the oceans. When a quake occurs and causes the ocean floor to drop, the displacement of the water creates a ripple in the entire column of seawater. That sudden movement translates to the ocean surface, causing waves that can eventually either threaten coastlines or dissipate. They can also propagate from underwater landslides and violent volcanic eruptions—and more rarely, by large space objects that drop into the oceans.

Tsunamis are often called "tidal waves," a misnomer frequently used in the popular press. True tidal waves are the periodic movement of water caused by the rise and fall of the tides (which, in turn, are controlled by the gravitational attraction of the Moon, and to a lesser extent, the Sun). It's the same for weather, too—tsunamis have no connection to weather and vice versa—no matter what the tabloids say.

A tsunami in the deep ocean is likely not to cause much of a stir—the extremely long wavelengths can measure hundreds of miles between the crests of the waves, making it seem like just a gently rolling sea. They travel slowly compared to seismic waves, typically around 311 miles (500 kilometers) per hour or more, and can span more than 62 miles (100 kilometers) from crest to crest.

The trouble comes when a tsunami wave reaches closer to a

coastline's shallow waters. The wave literally drags its belly along the bottom, causing the speed of the wave to decrease and the heights—sometimes up to 100 feet (30 meters)—to increase. It transfers its energy to the ever-decreasing water mass, causing even more destructive waves. As they reach the coastline, they often break like waves; but some do not break, rather surging like a massive, fast-moving high tide. Interestingly enough, much of the damage doesn't occur as the waves flood the low-lying areas, but as they recede back to the ocean.

Depending on the distance from the earthquake, the speeds and direction at which the waves strike the shore, the shape of the shoreline, and the density of the population, a tsunami can either be a nuisance or a megadestructive hazard. For example, a shoreline with wide bays that narrow can act like a funnel, increasing the destructive powers of the waves; whereas a shoreline with an offshore coral reef can dissipate the tsunami's energy. Even a coastline at an angle to the propagating waves can experience a decrease in the destructive power of the massive waves.

Without Warning
Scientists realize the Pacific Ocean is the true hotbed of tsunamis. After the 1946 tsunami, a tsunami warning system for the Pacific Basin was developed in 1948 by the United States Coast and Geodetic Survey. It then became the Pacific Tsunami Warning Center; today it is stationed in Ewa Beach, Hawaii, and administered by the National Weather Service (under the National Oceanic and Atmospheric Administration [NOAA]).

Not all places have the advantage of a huge warning system like that in the Pacific. One of the reasons stems from statistics: there just aren't that many tsunamis occurring in the Indian, Atlantic,

> ## Simply a Seiche
>
> A *seiche* is actually sloshing water in a closed body—easily seen on a smaller scale when you push water in a bathtub in one direction and start a rhythmic movement back and forth. A seiche occurs naturally in bays, estuaries, lagoons, and even swimming pools. On occasion, tsunamis can produce seiches as a result of a local area's topographic peculiarities. For instance, the tsunami that hit the Hawaiian Islands in 1946 had waves at fifteen-minute intervals between wave fronts (or "ripple" in the waves). The natural resonant period—or sloshing back and forth—of Hilo Bay at the island of Hawaii was about a half hour. That meant that every second wave was in phase with the motion of Hilo Bay, creating an even larger seiche than usual. As a result, Hilo suffered worse damage than any other place in Hawaii—with the tsunami/seiche reaching a height of almost 50 feet (15 meters) in the most-affected areas.

and Southern oceans, to name the largest oceans. Because of this lack, there just doesn't seem to be a call for such an extensive warning system. But that doesn't mean a tsunami caused by an earthquake can't happen in the wrong place—a painful lesson we all learned in December 2004.

Take the long chain of islands collectively known as Sumatra that lies within the borders of Indonesia. The islands also lie not far from the Sunda trench, a lengthy subduction zone speckled with volcanoes just inland along its length. It is here that the Indian plate slides under the Burma (or Sunda) plate in the eastern Indian Ocean—a crustal chunk that holds Indonesia, Malaysia, and most of Southeast Asia. Across the Indian Ocean from Sumatra lie India and Sri Lanka; nearby are Thailand and sundry lands such as Nicobar and the Andaman Islands. The Sunda fault was known to be under an accumulating strain—but fault lines are known to give in with a few minor to moderate earthquakes.

The day after Christmas, December 26, 2004, the strain became unbearable for the land to hold. Without warning, a giant quake measuring around Moment Magnitude M_W 9.3 occurred, the second largest ever recorded. The result was catastrophic, as the 60-mile (100-kilometer) rupture caused a record 750 miles (1,200 kilometers) of the Sundra fault to slip.

But the Sumatra–Andaman earthquake, as it is now officially called, was not the major factor affecting human lives. It was the massive tsunami generated by the huge earthquake. The fault moved about 50 feet (15 meters), causing the seafloor to rise just over 15 feet (5 meters). In response to the fault movement, ocean water was displaced, essentially dragging the water down, creating a ripple on the surface. Like dropping a stone in a smooth pond, the waves spread out—tall tsunami waves racing toward nearby coastlines.

There was no warning—as there was no real tsunami warning system in place. Coastlines were devastated—from Indonesia, Sri Lanka, India, Thailand, and other countries—with waves up to 100 feet (30 meters) high. With most of the population living along the coastlines, the destruction was immense. After the December quake, the area experienced shakings from aftershocks—not a surprise to scientists—with swarms occurring often in blocks of about a week. It is estimated that 230,000 people drowned from the tsunamis alone—an estimate that many people fear is much lower than the reality. The devastation was so extensive, we may never know the true number of people lost.

What caused so much death and destruction in the December 2004 Indian Ocean tsunamis? Probably more than anything it was the magnitude of the quake—around Moment Magnitude M_W

9.3—and the deep, subsequent tsunami ripples. The quake's epicenter was no surprise. It was situated near the location of the meeting point of the Australian, Indian, and Burmese plates, a region of compression as the Australian plate is rotating counterclockwise into the Indian plate. What most people don't realize is that this region also experienced a second extreme jolt in March 2005, the Sunda fault shifting just southeast of the December rupture. The resulting quake was not as violent, and the resulting tsunami caused "minor" waves, with "only" (as the media put it) an estimated 4,000 resulting deaths, mostly in Sumatra.

Amazingly, this was not the first time a tsunami struck the region. One other famous wave occurred, for a much different reason. In 1883, the violent volcanic eruption at Krakatau led to a three-foot surge of water in Sri Lanka. What was the big difference? With fewer people populating the coastlines and a shorter wall of water, there was much less damage. And in the last century, about seven tsunamis have been reported, set off by earthquakes near Indonesia, Pakistan, and one at the Bay of Bengal—but none large enough to warrant a warning system.

Still, some people doubt if fewer people would have died with a warning system in place. After all, no one would really expect such large waves in this region. In fact, there has never been any record of earthquakes causing giant tsunamis in northwestern Sumatra, only moderate ones. Would anyone have known what to do? With the December 2004 quake fresh in everyone's memory, most of the countries around the Indian Ocean now carry out tsunami drills, hoping that if another giant tsunami crashes on their shores, this time everyone will know what to do.

Translating the Sumatran 2004 Earthquake

After a catastrophe in the natural world, scientists often race to publish scientific observations and results of the tragedies. And after the December 26, 2004, Sumatran earthquake hit, there was a chance to do just that—publish hundreds of papers on the subject. But Thorne Lay, professor of earth and planetary sciences at the University of California, Santa Cruz, decided to change the way this earthquake was viewed. He didn't want the tragedy to be treated in the usual way. Lay and other scientists finally decided on a collective effort from the seismological community—to use the data to produce a single account of what truly happened on that horrible day.

The account was published in a 2005 issue of *Science*. The resulting papers were coauthored by 40 researchers from 23 institutes and universities located in seven countries. The results were startling: it was the largest such earthquake anywhere in 40 years, with the strain energy released equaling all other earthquakes combined over the past 15 years. Aftershocks in the Andaman Sea followed along shallow faults—the largest swarm of magnitude 5.0 or greater aftershocks ever observed.

The researchers also discovered that the vibrations from the quake were immense in terms of global effects. In Sri Lanka, about 600 miles (1,000 kilometers) from the focus of the quake, the ground shook with amplitudes greater than 3.5 inches (9 centimeters) after the first compressional waves arrived. Even more impressive, the long-period surface waves traveled around the entire Earth, the motion eventually causing the surface to move just under a half inch (1 centimeter) around the world. Overall, the quake lasted nine minutes, and ruptured at least 95,526 square miles (250,000 square kilometers) of oceanic crust (with a certain section moving at least 30 feet [10 meters] across the fault surface).

The effects didn't just stop after the Sumatra quake. The shaking also created a kind of resonance in the Earth, like the bonging of a huge bell. This included one form of vibration called a "breathing mode," in which the Earth's surface expands and contracts from a massive shaking. This time was no different—such vibrations were observed for weeks after the initial quake.

Other such studies have followed, including one conducted by the Institut de Physique du Globe de Paris (IPGP) called the Sumatra-Andaman Great Earthquake Research Initiative. Scientists had discovered that the fault line along the Andaman trench—a stretch of the Indo-Australian plate that sinks (subducts) under the Eurasian plate—ruptured. It started near Simeulue Island, progressed slowly for a minute, and then grew in intensity, with the greatest amount of slippage along the fault near Sumatra's northern edge. It then progressed northward, past the Nicobar Islands, then twisted clockwise toward the Andaman Islands, unzipping the crust with ease. Scientists weren't surprised to hear about the tear in the southern part of the rupture just off northern Sumatra—it's a young, fluid area of the plate that is more active. But when the northern, older part unzipped, it was a surprise. Even more of a surprise was the possible channeling of the earthquake rupture in a narrow zone between the trench and another, previously little-known crustal boundary—a funneling that could account for the huge December event.

This is definitely not the end of the activity in this part of the world. There is a chance that stress is accumulating along this boundary, which will lead to another major earthquake in the near future—from years to decades. In the meantime, the shaking continues, albeit without the accompanying destructive tsunamis: keeping with the idea of earthquake swarms, the March 28, 2005, earthquake off Sumatra that measured a Moment Magnitude M_w 8.6 may have been triggered by the December quake—and was the second largest in 40 years. A quake on December 1, 2006, occurred at the northern part of Sumatra, about 310 miles (500 kilometers) from Banda Aceh, a Moment Magnitude M_w 6.3 that caused no damage or casualties. But the December 18, 2006, quake killed 7 and wounded more than 100—a Moment Magnitude M_w 5.7 just before dawn, southwest of Banda Aceh, with many lesser-magnitude aftershocks.

The "What-if" Factor in the Atlantic Ocean

The tragedy in the Indian Ocean's December 2004 tsunami was made even more poignant by the fact that there was no tsunami warning system in place, as there is in the Pacific Ocean. Such a system probably would not have saved as many people in Sumatra, the lands closest to the immediate effects of the earthquake; but others along the numerous affected coastlines would have benefited; as many people as possible could have been moved inland in the several hours before the tsunami struck. The actual wave traveled inward a few miles; even a warning to get as many people as possible outside that tsunami red zone would have saved countless lives. Today a warning system is being constructed in the Indian Ocean, hastily put into effect after the December earthquake, with the first early warning buoys deployed in the ocean between Thailand and Sri Lanka.

Not only is the Indian Ocean ripe for such occurrences, so is the Atlantic Ocean. Here, the spreading movement of the Mid-Atlantic Ridge, a long underwater chain of mountains that cuts the Atlantic Ocean seafloor, is one reason for quaking; the other is the moving of the smaller Caribbean plate. The vulnerable coastlines are many, and include Africa, western Europe, South America—and the most populated coastline in the United States, the East Coast, with metropolitan areas that include New York City, Boston, and Norfolk. It's true that there is only about a 10 percent chance of a tsunami striking in the Atlantic Ocean, but 10 percent is enough—a one in ten chance. In fact, scientists believe that a severe tsunami strikes the western Atlantic (translation: the East Coast) once every 100 years or so.

Twenty-four tsunamis have caused damage in the United States and its territories during the last 200 years. This may seem

The ubiquitous geyser at Yellowstone National Park—Old Faithful—and evidence of a possible hotspot underneath the park. *Courtesy of USGS.*

like cause for concern, but to most people it is not. Since 1946, six tsunamis have killed more than 350 people and caused at least a half-billion dollars in damages—but most of these effects have been felt only in the states of Hawaii and Alaska and along the West Coast, not the East Coast.

Thus, a Pacific-type warning system does not exist in the Atlantic. But what would happen if a huge quake struck off the coast—say a deep-ocean magnitude 7.4 earthquake just off Iceland along the moving Mid-Atlantic Ridge? There are no buoys to record such a change in water levels off the shore of Iceland—and just a few toward the North Sea. And the North Sea is definitely not near New York City.

The results could be disastrous, depending on the time, location, and magnitude of the quake. Giant walls of water would hit

The Halemaumau fire pit within the Kilauea crater at Volcanoes National Park–all on the Big Island of Hawaii. *Courtesy of NOAA.*

and collapse houses hugging the coastline of Maine; then, like a domino effect, waves would strike down the eastern seaboard to Florida. As waters reached inside the harbors of Boston and New York City, windows of skyscrapers and smaller structures would blow out and ships would be tossed onshore. By the time the tsunamis reached the Outer Banks along North Carolina, the sands would rapidly change, shifted by seismic and ocean waves. As the tsunami's effect reached northern Florida, it would probably dissipate—but not enough to stop the water from running over parts of the lowest coastlines. And without a tsunami warning system in place, many lives would be lost.

This is not science fiction, either. A tsunami hit the East Coast of North America in 1755—around eight hours after a major quake shook the coast of Lisbon, Portugal. Florida was affected, but the sparse population in the state at that time saved it from a major catastrophe (the same would not be said about today's Florida coastline); the wave height was also greater in the Caribbean, ranging from 10 to 15 feet (3 to 5 meters). Another noteworthy tsunami in North America arrived after the 1929 Grand Banks earthquake near Newfoundland, which measured a magnitude 7.3 and reached as far south as South Carolina. In 1867, 20-foot (6-meter) and 30-foot (9-meter) waves struck parts of the U.S. Virgin Islands after a quake. This area is home to a much larger population—along with plenty of visiting cruise ships and oil tankers that frequent the harbors. A similar tsunami would be devastating, causing up to a billion dollars or more in damage.

Florida's Rogue Wave

Just as there are rogue earthquakes, there are rogue tsunamis. One such wave hit Daytona, Florida, on July 3, 1992, just before midnight. According to reports, the wave was 18 feet (5.5 meters) tall. It reportedly damaged around a hundred cars, and about 20 people were injured by the wave out of nowhere. No one knows where the wave originated, though it was first thought to be from an underwater landslide. Later reports blamed a series of thunderstorms that originated along the Georgia coast. A tide gauge picked up a large southerly-moving wave near Saint Augustine around 50 minutes before the wave reached Daytona—meaning this rough wave may have been propagated not by a quake or slide, but by weather.

Tsunamis without Warning

Other tsunamis have occurred seemingly out of nowhere—without any warning and under stranger-than-usual circumstances. In particular, a tsunami can suddenly occur along a stretch of marine coastline—in association with a steep continental shelf or volcano—or on the shores of a lake. For example, in 1792, the Unzen volcano in Japan collapsed, dumping a huge chunk of rock and debris into a shallow inland sea, generating a sea wave that killed close to 15,000 people along nearby coastlines. In the early morning hours of March 13, 1888, about 1.2 cubic miles (5 cubic kilometers) of the Ritter Island volcano fell into the sea northeast of Papua New Guinea. This created the largest lateral collapse of a volcanic island in historic time, and caused a tsunami tens of feet high that struck adjacent shorelines.

Bays, sounds, and lakes are not exempt, either. Scientists have discovered that a major earthquake that struck between 900 and 930 AD lifted central Puget Sound in Washington by about 20 feet (6.5 meters)—and created a massive tsunami. Using physical data from around the sound, scientists have constructed a computer simulation of the event—showing that the wave reached heights of more than 10 feet (3 meters) along today's Seattle waterfront. The region was flooded again after the April 1949 tsunami, but this time, it did not occur right after a quake. On April 16, three days after a magnitude 7.1 quake struck the region, a landside occurred when a 400-foot (122-meter) cliff collapsed and slid into Puget Sound. The water receded some 25 feet (8 meters) from the normal high tide, then rushed in with an 8-foot (2.5-meter) wave, slamming boats, docks, and waterfront buildings along Salmon Beach.

Lakes also experience tsunamis, with or without a nearby earthquake. In 1980, after the May 18 eruption of Mount Saint Helens, a massive tsunami occurred in Spirit Lake as the volcanic activity blasted out the side of the volcano. The sliding north face slammed into the west arm of Spirit Lake, raising the surface an estimated 207 feet (63 meters). The resulting tsunami ran through the lake, raising the water as high as 820 feet (250 meters) above the previous lake level; the east arm was also affected, the tsunami wave reaching about 740 feet (226 meters) above the old level. As the wave propagated across the water, it scoured the shores of the lake. The resulting mix of water, logs, boulders, and debris ended up along the shore and on the landslide itself.

Lake Tahoe is another example: this resort area in California has a system of underlying faults—some of which have ruptured in the past. Three of the main features are normal faults, in which one block of crust moves abruptly downward during a major quake. Such movements have occurred at the lake in the past, resulting in a tsunami that scientists estimate could have been 30 feet (10 meters) high. A double whammy occurs in Lake Tahoe, too—a landslide along the lakefront can also create a tsunami. A massive landslide was detected in 2002 along the lake bottom. It is thought that the slide occurred about 60,000 years ago at McKinney Bay on Tahoe's western shore—and probably produced huge tsunami waves.

In truth, no matter where a tsunami strikes, it's not the waves that are the real problem—it's the population and its inability to get out of the way. Once the shaking is detected (at least currently in the Pacific Ocean and hopefully in other oceans in the future), a warning of a potential tsunami goes out, giving most coastal places several hours' notice (though people living right near a massive earthquake would have only minutes to respond). The biggest problem then becomes education: scientists note that unlike people living around the Pacific Ocean, people living around the Indian Ocean are not taught about the possibilities of tsunamis. Most countries around the "Ring of Fire" teach coastal dwellers *not* to run out to catch fish or pick up shells when the ocean water quickly recedes from the shore—or if they feel tremors. They are taught to run the other way, to higher ground.

Dam Tsunami

While most landslides aren't predictable, they can be frequent. For example, after the Grand Coulee Dam was built on the Columbia River in eastern Washington, the resulting Lake Roosevelt generated numerous tsunamis. All the waves were from non-earthquake-generated landslides along the lake, caused by loose underwater sediments. One of the first occurred on April 8, 1944, when a four- to five-million-cubic-yard (two- to three-million-cubic-meter) landslide created a 30-foot (10-meter) wave, around 98 miles (158 kilometers) above the dam. Dozens more occurred in the 1940s and 1950s, including the October 13, 1952, slide, again 98 miles (158 kilometers) up from the dam. The wave broke tugboats and barges loose from their moorings around 6 miles (10 kilometers) away, while logs and other debris landed in a huge area above lake level. (Most of the lake's landslides abated after 1953; currently, the levels of the lake are down because of a drought, making the chance for landslides definitely lower.)

Tsunamis from Space?

Could an asteroid or comet strike the Earth and cause movement? According to most astronomers, any asteroid or comet greater than about 330 feet (100 meters) in diameter would have more of a localized than a global effect. But the local effect would be catastrophic—especially if it occurred in the shallow waters off a coastline. First it would smash into the surface; then it would create an earthquake-like effect, sending seismic waves through the Earth. The ocean would also respond, absorbing much of the impact energy and possibly creating a somewhat moderate tsunami from the ripple effect. Beyond that, the area would experience short-term changes in the weather, as the natural hot cannonball sprayed ocean water and steam into the atmosphere.

But what about a slightly larger asteroid or comet?

One of the more drastic ways to propagate a giant tsunami is often relegated to the world of fiction: a strike in the oceans by a huge meteoroid, comet, or asteroid. Scientists have long known that large space objects have hit the Earth, as there are about just shy of 150 impact craters found on land. But it's the possible impacts in the oceans that cause the most consternation: with a large enough meteoroid, the resulting ocean ripple could wipe out whole coastlines. No one knows where such a strike would occur. Unlike most quakes, which occur along fault lines near tectonic plates, an asteroid, comet, or meteoroid doesn't have such restrictions.

They can strike anywhere—land or sea. A land strike may be worse, as a great deal of debris would be tossed into the atmosphere; whereas the direct and indirect effects might be less hazardous to humans. But of course, in the oceans, the larger the space object, the bigger the splash. One study estimates that an asteroid 164 feet (50 meters) in diameter would raise the ocean wave height 0.39 feet, or just under a half foot (0.12 meters) above sea level at a point 62 miles (100 kilometers) from the impact, creating a localized effect. The giant waves would be propagated from larger objects: a 656-foot (200-meter) asteroid would raise the height by 10 feet (3 meters); and a rock just over a half mile (1 kilometer) would raise the sea height by 230 feet (70 meters).

One recent impact scenario was proposed by scientists Steven Ward and Erik Asphaug at the University of California, Santa Cruz: according to the most recent orbital estimates, the asteroid known as 1950 D, a meteoroid around two-thirds of a mile (around 1 kilometer) in diameter, is set to swing close to the Earth on March 16, 2880. None of us will be here then, and it may not truly strike our planet—the probability of impact is only about 0.3 percent.

The researchers took that rock, put all the data into a computer simulation that would make it slam into the Atlantic Ocean, and determined just what would happen: the impact would generate tsunamis as high as 400 feet (122 kilometers) and come in many waves. The resulting ripples would inundate the coastlines of western Europe and Africa; on the other side of the Atlantic, along the East Coast of the United States, waves would flood deep into the states, reaching the Appalachian Mountains, flooding New York City, and making upstate New York prime beachfront property. (Though there is some "good news": the gradual slopes of the British Isles' and Florida's continental shelves may reflect much of the tsunamis back into the ocean.) Secondary effects would include submarine landslides, which could, in turn, create more tsunamis. And forget Iceland and small islands like the Azores—they would no doubt become legends, like Atlantis.

And this is not a drill. Scientists have found evidence of hundreds of such space-induced tsunamis that have occurred since the age of the dinosaurs—more than 65 million years ago. For example, the Eltanin impact by a large asteroid (estimated to have been between just over a half mile to 2.4 miles [1 to 4 kilometers] in diameter) struck just off the coast of Chile some 2.15 million years ago. No crater was formed on the ocean floor (and no extinction occurred), but it caused a massive tsunami, with wave heights in some locations greater than 200 feet (61 meters). It's happened before—and no one knows what the chances are of it happening again.

LANDSLIDES

Slides and Shakes

The wall of mud, rock, and vegetation seemed to move as if it were alive. Around 10:30 A.M. local time on February 17, 2006, in Barangay Guinsaugon, municipality of Saint Bernard, Southern Leyte Province, Philippines, the nearby mountainside came sliding down at speeds reaching 62 to 250 miles (100 to 140 kilometers) per hour. Caught without warning, the villagers had little time to respond. As the landslide slowed and reached equilibrium, it came to rest at the bottom of the mountain. It covered the entire village—a mass of material with an estimated volume of about 70 million cubic feet (20 million cubic meters). It was the worst landslide in the region's recorded history, with around 122 people dying that day in a few short hours and more than 1,328 people missing. The slide, officially called a rockslide-debris avalanche, was one of the largest such events of the twentieth and twenty-first centuries.

The scientists who reached the scene not long after the event believe the Barangay Guinsaugon slide, as it is now called, had multiple causes. Rains saturated the hillside that month, with about 27 inches (68.3 centimeters) recorded at a rainfall station 4 miles (7 kilometers) from the slide site. Next, the slide had to have a trigger. At first, it seemed as if a magnitude 2.6 earthquake striking the region just before the slide was the culprit. But the quake occurred 16 miles (25 kilometers) west of the landslide area, and it seemed too weak to cause such a catastrophic slide. Interviewing local residents added more information: two months before the disaster, the residents felt an earthquake and noticed cracks in the ground; and two days before the landslide, the river between the base of the scarp and the town dried up. The water became the center of focus—and now scientists believe the river and rainfall may have seeped into the fractures. This essentially "lubricated" the area where the rock broke away from the mountain, sending a massive chunk of material to the town below.

In many parts of the world, landslides are nothing new. Asia is at the top of the list, known for having the most landslides in the world. Interestingly enough, it is not the region with the most deaths and injuries from landslides; that distinction goes to the Americas.

And not all of the slides are due to earthquakes. For example, as Hurricane Stan moved across Central America in 2005, it dropped torrential rains on an already saturated land. The storm triggered a multitude of landslides of all sizes across the country. One particularly hard-hit area was the small Mayan village of Panabáj, Guatemala, on the flanks of the volcano Atitlán in October 2005: the loosely packed volcanic sands and clay soil created a slurry of mud and debris, killing more than 500 people, most of them buried alive in their homes. Some tried but were unable to outrun the flow—as the slide traveled an estimated 35 miles (50 kilometers) per hour. As the flow settled over the city, the volcanic slurry hardened—as a USGS scientist said, "setting up like concrete." Panabáj was not alone. According to Guatemala's disaster management division, two other mudflows killed a total of almost 1,000 people as the hurricane rolled through, with hundreds reported missing. Why do people still live here? The draw in this region is the rich volcanic soils perfect for coffee and other crops. Even though evacuation reports were issued for the hurricane, many mayors of the various stricken towns reported that they had received no reports—thus, they say, the great death toll.

But there have been hundreds, if not thousands, of other slides propagated from the shaking of the ground. On October 8, 2005, a devastating earthquake struck Kashmir, a region shared by India and Pakistan, with most of the damage taking place on the Pakistan-governed side. It measured magnitude 7.6, but it didn't stop there. Two hours after the major event, the area was struck by a dozen major aftershocks, one measuring a whopping magnitude 6.2. The quakes caused massive collapse of buildings and the region's infrastructure, such as it was; and more than 74,000 people died, not only in the collapse of buildings but in the landslides caused by the shaking.

Satellite images taken after the quake allowed scientists to see the extent of landslide damage. Communities in the northern Pakistan region of the Kashmir often build their mostly mud structures along the narrow, steep canyons—areas that act like a funnel, carrying loose soils from landslides. During the October event, many of the quake-caused slides coalesced along these river valleys, numbering into the thousands, both large and small. One slide that measured about a mile (2 kilometers) long and just over a half mile (1.5 kilometers) wide blocked two rivers, forming a natural dam in an area that had previously been dry. Scientists know that such a new lake could lead to more destruction—especially if another large quake shakes the loose material holding back the river water.

Landslides of All Types
When most people hear the words "mass movement of Earth chunks," they think of landslides—the often rapid displacement of rock and/or soil. In general, the slides generally move by falling, sliding, or flowing, or a combination of all three, depending on the

location, size, and conditions that precipitate the slide. As most of us know, slides have a "downward" component—the materials moving down a slope or incline. But what most people forget is the "outward" component. In many cases, the debris, rock, soil, and water does not just fall, it essentially oozes out from the original source—a factor that is very destructive for most landslide-prone regions.

Why do these earth movements form? They are caused by earthquakes, excessive rains, and changes in vegetation; other major factors include gravity, the shape of the slope, certain soils (such as clays that expand when saturated or shaken by an earthquake), and even sudden movements that literally "shake" the soils off the slopes. And the most disturbing fact about such landslides is that they can happen almost anywhere—at almost any time.

Are there different types of landslides? Yes, there are rockfalls that can cause excessive damage; and debris falls and rock slides—both of which are among the most destructive of mass movements—and are most often triggered by excessive rain or melting snow, or even earthquakes. And even slow-moving slumps—masses of soil, rock, or rock fragments that most often move slowly along a curved rotational surface—can cause damage. They can sometimes be seen along highways in which graded soil along the roadsides is a little too steep and are often triggered by excessive rains.

Why are debris avalanches so destructive? Debris avalanches occur after the complete collapse of a mountain slope. These very high-velocity flows are huge, carrying tons of rock, soil, and debris down a slope—traveling for considerable distances even along a relatively gentle slope. They are most often triggered by larger events, such as volcanic eruptions or an earthquake.

The most disturbing part about slides is, of course, the loss of life. But another disturbing thing about these moving masses of earth is their lack of "popularity"—unlike earthquakes, floods, tsunamis, and volcanic eruptions, landslide events are one of the most overlooked natural hazards in the world. But they are extreme movements that seem to occur out of nowhere and at any time—destroying roads, houses, and people, and creating havoc for rescuers searching for survivors.

To highlight the problem, in January 2006, researchers from the United Nations University presented a report to international organizations and diplomats explaining the tragedy of landslides. Their report examined nearly 500 historical global landslides from 1903 to 2004, showing that the events caused billions of dollars in damage and affected, displaced, or killed more than 10 million people. And as the world population increases, scientists fear the numbers will only rise.

Vacation Slides

Landslides occur all over the world, and in the United States alone, they cause almost two billion dollars in damage each year. For those of you who have never heard of a landslide in the United States, don't be surprised—as I explained above, they are not as "notorious" as earthquakes, floods, and sundry other natural hazards.

But they have occurred in some surprising places. One of the most notorious places for slides is the Yosemite National Park—most often caused by saturating rains or by the occasional shaking of loose rock along steep cliffs. And sometimes you can't even get away from such slides on vacations abroad: in 2005, tourists were stranded at the Incan ruins at Machu Picchu when a landside blocked the only path out of the World Heritage cultural site.

Megaslide of the Future?

Scientists believe that one day there may be a quake and resulting underwater slide that will rival all other slides. One place cited for such a mega-event is Hawaii, in which, after an eruption, a huge chunk of the Mauna Loa volcano will break off, creating a mega tsunami.

The reason for concern was recently found: a huge unstable part of the volcano that could easily slip in the right conditions. Just before an eruption, parts of the volcano could expand, putting additional stress on certain flanks of the mountain. Since volcanoes are notorious for their tremors, especially before a significant eruption, the spewing lava would be the least of everyone's worries. Dozens of cubic miles of material could run down the side of the slopes, heading for the water. Once the volcanic rock and dust fell into the water, it would not only create

A tsunami wave striking Hilo, Hawaii's Pier 1 in 1946. The man seen in this picture was eventually identified as Antone "Tony" Correa Aguiar, a wharf foreman who was severing a cable to the *Brigham Victory*, a cargo vessel loaded with ammunition and explosives—thus averting catastrophe, but losing his own life in the process. *Courtesy of NOAA.*

steam, but like dropping a rock in a pond, would displace water. From there, the huge waves would spread out, some scientists say, as tall as a 10-story building. In less than an hour, the wave would hit the main island, inundating Honolulu, bringing with it not only seawater, but chunks of coral, sand, and rock inland at least 10 miles (16 kilometers)—maybe more.

Is this something for residents of and travelers to "paradise" to worry about? Like many unpredictable things in science, no one knows if this would happen in 100 or even in 10,000 years. But what scientists do know is that it has happened before, most often when the sea levels are high and the Earth's temperatures are warm—and we are now in such a period.

But there are over fifteen giant landslides surrounding the Hawaiian Islands—all among the largest slides known to exist on the Earth. These underwater slides are thought to have occurred within the past 4 million years, with the youngest only about 100,000 years old. In the past, when the slides moved, large earthquakes were often generated, huge chunks of the island's landmass slid into the ocean, and large tsunamis crashed into the coastline. Some scientists believe that not only does Mauna Loa have the potential to cause a problem, but that other large blocks of Hawaii are also beginning to slide. If such an underwater slide were to occur, the potential for the loss of life, property, and resources would be enormous.

CHAPTER 9

Bizarre Quakes out of Nowhere

A bad earthquake at once destroys our oldest associations: the earth, the very emblem of solidity, has moved beneath our feet like a thin crust over fluid.

—Charles Darwin, from The Voyage of the Beagle

No matter where an earthquake strikes—from the quake-prone boundaries to the wrong places—there are some aspects of the shaking that still boggle the mind. Some of these "quirks of a quake" studies are in their infancy, and some of the quirks have yet to be explained. And many of them can lead to quakes in all the wrong places.

QUAKE TRIGGERS
How to Remotely Trigger a Quake
One strange occurrence in the quake world is called an "earthquake storm"—a term used by the media to describe what is scientifically called a *remotely triggered earthquake,* an event much different than just a plain old earthquake. It is believed that these storms occur along specific faults or within the same tectonic

plate. As an earthquake strikes, the released stresses don't just disappear, but reappear in the form of a "storm of quakes." Depending on the region—and that means rock, soil, and fault lines—such stress can be redistributed to another region of the fault, which sets off another quake. These storms are like the proverbial thunderstorm marching across the flat plains of Kansas, constantly rumbling as it moves along—but in this case, it's all about the Earth's crust, not the atmosphere.

What's the difference between these remotely triggered earthquakes and aftershocks? Aftershocks occur after all major earthquakes; in an earthquake storm, large independent quakes are triggered over long distances and extended periods of time. In fact, remotely triggered quakes act like dominoes, with large earthquakes activating other quakes from a considerable distance—an effect first observed in volcanic regions.

One of the greatest controversies surrounding the remotely triggered events is the distances involved. Some scientists scoff at the thought of such events traveling over great distances. Contrary to the "butterfly effect"—a long-controversial idea in chaos theory that states that a butterfly flapping its wings in Brazil leads to a tornado in Kansas—a remotely triggered earthquake to many scientists does not mean that seismic activity in California will trigger quakes in Turkey; they believe the distances are too great and the variables too numerous to prove such an event.

But there are those scientists who believe there is a long-distance connection. For example, Susan Hough, a geologist at the USGS, proposes that a violent earthquake in New England in 1755 may have been triggered by the great Lisbon earthquake that occurred only 17 days before; and even the New Madrid quakes of 1811–1812 show evidence of remotely triggered quakes.

The physics behind such a chain of earthquakes is complex. But those who do advocate the possibility state that most of these earthquake zones are already near failure. One push triggers a cascade effect, starting the chain of quakes. According to Susan Hough, remotely triggered earthquake shaking is similar to the effect of "shaking a soda can": as the pressure from the earthquake waves raises the pressure of the underground fluids, more earthquakes are triggered. Some scientists advocate other theories, such as when the quake occurs at a distance, the region is either shifted by the initial movement of the crust, or the seismic waves shake things enough to cause the other quakes to occur.

Though scientists don't agree on a mechanism, many can't deny some of the physical evidence of remotely triggered quakes. The example often cited is what happened after the magnitude 7.3 Landers quake that struck California in 1992, a shaking that killed 171 people and caused more than a million dollars in damage.

The story starts with Parkfield, California—once referred to as "the earthquake capital of America"—where Ross Stein, a professor at Boston University, and others believed a major quake would hit. After all, quakes seemed to occur there once every 22 years; to catch the "next" quake, scientists set up a complex network of instruments by 1987 and waited. But it seems the wait was in the wrong place, because the next big quake in that region occurred in 1992 in the small desert town of Landers 250 miles (400 kilometers) away. It was one of the largest quakes in 40 years, shaking apart the community and creating the largest amount of ground movement in a California quake in over 100 years.

But the story didn't end there. About three hours after the Landers quake, a huge aftershock struck Big Bear, a town about

25 miles (40 kilometers) away. Stein and his team had already been working on a computer model that could help them study the relationship between earthquakes. They used the data collected during the Landers/Big Bear events to create a model to predict where the stress from Landers would have been transferred. It was the evidence scientists had been looking for—signs that the Landers quake may have triggered the Big Bear shaking—or, in other words, there was a transfer of stress along a fault. It appeared that as the quake released its seismic energy, it redistributed it to the west—right at Big Bear, a major quake causing an aftershock in a distant place.

Coming Unzipped
Landers was only one example—scientists needed more evidence to prove a major quake could trigger another over longer periods and far from the original quake. An area on the other side of the world soon became a contender—the North Anatolian fault, one of the most earthquake-prone faults in the world, which runs north of Turkey. Scientists studying the fault line noticed that there were four westward-migrating earthquakes that caused 450 miles (725 kilometers) of the fault to rupture.

It was as if this area of the crust were coming unzipped: the first quake occurred in 1939 in Erzincan, eastern Turkey; in 1942, Tokat, a city to the west, was next; more still occurred in a western migration path in 1943, 1944, and 1957; and, by 1967, the last quake occurred near Adapazari. The sequence exhibited a trend: quake, quiescent time, quake, and so on, migrating in a western direction. Overall, the entire line of quakes killed tens of thousands of people and destroyed countless structures.

The 1967 quake was not the last. Seismologist Geoffrey King, currently at the Institut de Physique du Globe de Paris, and his colleagues discovered that not only did these quakes occur along a "single" line, but it appeared that another city was in the line of sight. With the evidence, the scientists plotted previous quakes, showed that such a chain of quakes exists, and even announced that another quake in line would occur around Izmit and the Bay of Izmit, Turkey. Though the earthquake threat was published in the scientific literature and the popular press, it did not generate much interest or concern. It was, after all, only a prediction, and most people give little credence to such guesses, however educated.

The prediction did come true, but not right away. In August 1999, an earthquake struck Izmit. It only took 45 seconds to destroy much of the city; buildings even collapsed 62 miles (100 kilometers) away. There were warnings, all based on the "new" studies of remotely triggered quakes—but no one listened. It's estimated that more than 25,000 people lost their lives, and the damages rose into the millions.

Is there another remotely triggered quake in this region's future? Based on the data collected during the Izmit quake, scientists now believe the next quake in the chain will occur in the Sea of Marmara—a mere 12 miles (20 kilometers) from Istanbul, a city of some 10 million people. A quake the magnitude of the one in Izmit would cause immense damages and death. But, as with the last prediction, no one knows if a quake will occur in 100 or 10 years—or even within the next few years.

If the current theory of remotely triggered earthquakes does prove to be true, who knows what the potential destructiveness of one quake could be? Do such quakes occur often? How far do they reach toward the interior of a tectonic plate? And if there are more such remotely triggered quakes than was first believed, what does that mean in terms of earthquakes in the wrong places? With the study in its infancy, we have a long way to go in answering those questions.

Ancient "Earthquake Storms"?

Though the term "earthquake storms" may have been contrived by the media, it is thought that such multiple quakes may have wiped out entire regions in the past. Did such quake storms push the end of the Bronze Age? Did a chain of quakes—well known in parts of the vast Roman Empire—help to crumble one of the mightiest empires on Earth?

No one knows the answers to these questions yet, but there are some indications of other possible remotely triggered events in the past. For example, tucked into the records of many Mediterranean cultures are mentions of a great universal earthquake that seems to have struck the region after 360 AD. Physical evidence also exists—some coastlines experienced uplift in a relatively short period; and archaeologists have uncovered towns of that time in ruin, citing crushed skeletons, broken beams, and shattered pottery. All of this has been interpreted as the results of a major earthquake upheaval. But that wasn't all. Cities more than 932 miles (1,500 kilometers) apart seemed to show evidence of a massive quake, a distance almost as long as the Mediterranean itself—from Cyprus, North Africa, Crete, and on to southern Italy.

Was this truly one huge quake that shook each one of these cities? As more evidence was gathered, some of the quake dates just didn't add up. Now some scientists believe it was not just one single quake event, but an earthquake storm—with each quake triggering the next, down the line of cities.

BUNCH OF QUAKES
Swarming Quakes

As if remotely triggered quakes weren't enough, there are also quakes that occur in swarms. Earthquake swarms are often associated with volcanic regions or even with certain active faults along plate boundaries. But swarms can come out of nowhere, too.

For instance, Colorado isn't really a hotbed of quake activity—the USGS records an average of 8 to 10 quakes a year—and Segundo, Colorado, isn't exactly a typical earthquake-prone area. But in 2001, this region of the Raton Basin was treated to not one strong quake, but multiple tremors stretching over many weeks. The damage was minor, but it was enough to make residents wonder if they were suddenly in a growing rift valley. Such quakes, usually measuring from magnitude 2.0 to just over magnitude 4.0, are called earthquake swarms—a somewhat continuous series of quakes that occur over short periods. The verdict is still out, but the reasons for the swarm near Segundo may be a previously unknown fault, injection wells associated with a coal-bed methane operation nearby, or both.

In other areas, earthquake swarms seem to come out of nowhere, too. One such swarm struck Spokane, Washington, in 2001. Spokane's largest quake in a hundred years, it struck the area on June 25, measuring magnitude 3.7. By August 1, 37 minor earthquakes had been detected, with 6 hitting on July 31 alone; if the quakes detected by monitoring seismographs were added, there were more than 50 seismic events. The quakes in this earthquake swarm were shallow—only a mile or two deep—and very localized. No one truly knows why the swarm struck (it has since disappeared), especially since the area was short on seismic recording devices. Some geologists believe it's the inevitable direct effect from the nearby North American plate moving past the Pacific plate—with Washington and Oregon being squeezed in the middle. Yet another group of scientists has identified a new fault in the area—the Latah Creek fault—which may have been responsible.

Water Rights
It's been studied for more than a century, but not very intensely until recently: the response of groundwater to an earthquake, and a quake in response to groundwater. The water–quake connections—and there are probably many—remain controversial. But because there have been so many bizarre possible connections, natural water-pressure fluctuations and the shaking of the ground has become a whole new field of earthquake study. And since groundwater is everywhere under the surface, it adds another dimension to quakes occurring out of nowhere.

In a sort of strange twist, it appears that a quake in one place may cause a displacement of water in another place—almost like a watery form of a remotely triggered earthquake in reverse. How does it work? It has been documented that after certain earthquakes, the groundwater levels change and there is faster stream flow. According to many geologists, the tremors compress and stretch the Earth's crust, squeezing sediments like a sponge. This results in an outflow of water stored in the pore spaces within the sediment, causing water to flow faster and well water to rise.

But not everyone agrees. Some scientists don't believe that the earthquake shakes the fluids out of rocks and sediments—but the shaking does dislodge some of the biogeochemical glue that holds the sediments together from inside rock fractures. Like the goop in your drain before the plumber comes to unplug your sink, this biogeochemical mix is shaken, then washed away; and like your unclogged drain, the bathtub fills again—or in terms of nature, the wells rise. Others believe a seismic wave can cause the water to rise or lower as it passes, while still others believe the quake causes fractures in the ground, breaking the seals around wells.

Whatever the mechanism, water in wells often acts like a kind of seismogram, responding to the expansion and contraction of the surrounding rock material. For example, after the March 27, 1964, quake in Alaska, the waves traveled almost 3,000 miles (4,827 kilometers) in slightly more than seven minutes. Well water in Missouri rose and fell, and the water became muddy. Water in many of the wells fluctuated drastically, and while some levels returned to normal, some did not. Another example occurred in November 2002: well water near Fairbanks and the North Pole rose a few fractions of an inch (a fraction of a centimeter) in response to the magnitude 7.9 quake in Denali, Alaska. But it doesn't happen only in the backyard of the earthquake's epicenter—the distances involved are often huge. Those same wells around the North Pole rose again a few fractions of an inch in December 2004. The rise was in response to the giant Sumatra earthquake that caused the devastating tsunami, about 7,000 miles (11,263 kilometers) away—and is a good example of the crust's elasticity and how easily seismic waves travel through the Earth.

Changes in well water can even occur in response to smaller quakes. For example, in 1998 in northwestern Pennsylvania, a magnitude 5.2 earthquake gently shook the local towns. It also caused about 120 homes to lose water—each well going dry within three months of the quake.

As usual with science, there is another way of looking at water and quakes—does the water actually trigger a quake? Some scientists seem to think so, especially when it comes to aftershocks, often citing the example of Iceland in June 2000; it is thought that water played a major role in triggering aftershocks and even more quakes after two magnitude 6.5 quakes shook the country. From well data collected immediately after the event, the researchers speculated that well-level changes increased the fluid pressure in the rocks along other faults in the area—causing failure along those faults and leading to more quakes.

Even more curious is the possible connection between quakes and mountain snowmelt that recharges groundwater in some areas. For example, on Oregon's Mount Hood, scientists reported, earthquakes around this volcanic area seem to, on average, follow about 150 days after the snowmelt enters into the ground. Can this—or does this—occur on snowcapped mountains everywhere?

A Really Bad Year

According to a 2005 report from the USGS, the year 2004 was the deadliest earthquake year since the Renaissance Age more than 500 years ago. That was when, on January 23, 1556, an estimated magnitude 8.0 earthquake struck Shanxi, China, and an estimated 830,000 people died. The year 2004 was also the second-most fatal on record, with more than 230,000 deaths (and that's only an estimate) from the December 26 quake and subsequent tsunami in the Indian Ocean; during the rest of the year, only about 1,000 deaths had been reported due to earthquakes.

What were some of the highest-magnitude quakes in 2004? One was the Moment Magnitude M_w 9.3 that hit Banda Aceh, Indonesia; a magnitude 8.1 struck north of Macquarie Island, about 1,000 miles (1,600 kilometers) southwest of New Zealand, hitting about three days before the Indian Ocean quake and tsunami. In the United States that year, a magnitude 6.8 in southeastern Alaska was the largest quake.

A Case for Silence

One more recent discovery is the presence of slow, silent—or *aseismic*—earthquakes. In a seismic earthquake, seismic waves are produced from the shock of the event; in the case of aseismic quakes, they are too slow to cause the ground to shake, and thus are not considered a hazard. These silent earthquakes are usually associated with subduction zones, such as those found in Japan and Mexico, but also more recently in non-plate-boundary zones. Their source is unknown—as is their connection (if there is any) to spawning a large quake.

But no one is ready to ignore these quakes. In the Pacific Northwest, scientists have found such quakes—and it has caused concern. They realize that this part of the United States, along with British Columbia, was struck by a quake measuring approximately magnitude 9.0 in 1700; some believe there is an earthquake pattern and that another great quake will occur in 300 years—or literally, right now. In 2001, Canadian scientists discovered that silent quakes in this region occur around once every 14 months in almost clockwork-like fashion. Could changes in these silent quakes be indicative of a major quake in the making?

These silent quakes have also been found in association with an area far from a plate subduction boundary—the Hawaiian Islands. Recently, scientists examining the island of Hawaii measured four silent quake events using a dense network of Global Positioning System (GPS) stations. Each silent quake displaced the ground in the same place, with the GPS stations moving in the same direction about the same amount. The last aseimic event in 2005 was associated with swarms—about 60—of small seismic quakes measuring between Moment Magnitude M_w 2.0 and M_w 3.0. The researchers believe these swarms of "micro-earthquakes" are a definite sign that the silent tremors are adding to stress in the fault zone. And if the stress becomes great enough, they believe a major quake will occur. (Interestingly enough, the last aseismic earthquake struck Hawaii's Kilauea volcano in January 2005; in October 2006, Hawaii experienced a Moment Magnitude M_W 6.5 quake, the third-largest quake on the island and the largest since 1989.)

If all of these examples hold true, these movements may eventually be used as "silent alarms," indicating that a major quake is on the way—maybe even a quake measuring as high as Moment Magnitude M_W 8.0 and M_W 9.0. Currently, earthquake hazard forecasting using the silent temblors may work only in association with subduction zones, but scientists are hopeful that some may eventually be used as "quake predictors" around other fault regions.

Future Quakes: What's the Risk?

CHAPTER 10

Forecasting Earthly Motions

> We are becoming more vulnerable to natural disasters because of the trends of our society rather than those of nature. In other words, we are placing more property in harm's way. . . . In many ways, the trends seem paradoxical. After all, most natural disasters occur in areas of known high risk such as barrier islands, flood plains, and fault zones. Over time, one would expect that the costs of natural disasters would create economic pressure to encourage responsible land use in such areas. The long-term economic impact of low-probability, high-cost events such as earthquakes and hurricanes are not being incorporated into the planning and development of our societal infrastructure. Economic incentives for responsible land use have been stifled by legislated insurance rates and federal aid programs that effectively subsidize development in hazard-prone areas. And while there will always be great political pressure to provide economic relief after a disaster, there has been little interest in requiring predisaster mitigation.
>
> —From "Why the United States Is Becoming More Vulnerable to Natural Disasters," G. van der Vink, et al., EOS newsletter, November 1998

FORECAST PROBLEMS
Why We Don't Know
One of my professors once told me that forecasting earthquakes lies in the same category as weather prognostication. At this juncture in time, with the technology that we currently have at our disposal, we know "diddly-squat" (he was from the South)—and nature bats last all the time.

Why? No matter how many computers you use, no matter how much data you collect, at this time in our technology, there are too many variables that can change or shift a storm to the east, cut off or slow down the progress of a low pressure system, and even change the course of a hurricane.

Earthquake forecasting runs into the same barriers. Scientists know the "where" of quakes associated with seismic waves from volcanic activity or a small quake along a known earthquake-prone plate boundary—but it may be impossible ever to predict the "when." The number of variables is seemingly astronomical: Does the fault shift the rock and stop it from shifting again? How much does pressure from overlying rock affect the quake zone? Does the movement of the plate on one side of the Earth disrupt a fault on the other side? When it comes to weather, the variables that control the skies seem endless; but, unlike the weather, earthquakes and their variables occur out of sight and underground.

You add more problems to the mix when you discuss quakes along a non-plate boundary such as the New Madrid region or northern Europe—not only with the lack of data necessary for statistics, but with all the unknown variables that occur with such a quake. Studying quakes along plate boundaries is difficult enough. It will take many more decades of study, data collection, better ways of manipulating earthquake information, and knowledge of the physics to know why and how such "out of nowhere" earthly movements occur. And only then will scientists *maybe* begin to understand just how to forecast such quakes.

Reading Statistics and Physical Data

Earthquakes have been around since humans have—and no doubt, before and after historical records, they have been fair game for people with a propensity for prediction. To contemporary scientists, forecasting (they prefer this word over *prediction*) earthquakes is based on statistics (how many times in a certain spot, how many large ones in a certain decade, and so on) and physical evidence (foreshocks, an increase in tremors, and so on). And the field has a long way to go.

Statistics is probably the most-used method in earthquake forecasting. Scientists rely on past data to predict future quakes, a statistical field that is not at all precise, mainly because of the lack of data and the variability between quake events. It's true that earthquake predictions based on statistics take on a "broad brush" view of earthquakes, but for now, it's probably the best approach.

For example, in the past two decades, instead of predicting specific events over short time scales—from hours to days—scientists started to concentrate their forecasting efforts on the probability of earthquakes over longer periods. For example, in 1988, based on past quakes in the Santa Cruz Mountain region, the USGS stated that there was a "high probability" a major earthquake would hit the region sometime in the next 30 years. A year later, the Loma Prieta quake shook the region.

But this doesn't always work. For instance, as stated previously, in 1983, scientists at the USGS forecast that a moderate earthquake would strike near Parkfield, California. The prediction was based on the observation that earthquakes with magnitudes of about 6.0 had occurred there in 1857, 1881, 1901, 1922, 1934, and 1966. Statistically speaking, this averages out to be about once every 22 years. Thus, the scientists stated that a quake would strike the region in 1988—give or take five years. The area was extensively monitored, and everyone from scientists and government officials to the residents held a collective breath. But they all had to start breathing again by 1993—nothing happened and the prediction was cancelled.

Other quake forecasts have been based more on physical evidence and not as much on statistics. Take, for example, the often-cited—and successful—scientifically based earthquake forecast that occurred on February 4, 1975, in Haicheng, China. The prediction was based on data collected months before the quake: changes in groundwater levels and ground elevation, dozens of foreshocks, and even reports of peculiar animal behavior. All this led to the China State Seismological Bureau ordering an evacuation of the region the day before a Moment Magnitude M_W 7.3 quake struck. In 1999, another successful prediction was made of the November 29 earthquake—a Moment Magnitude M_W 5.4 Gushan-Pianling quake in Haicheng and Xiuyan, China—made a week before the quake. Both predictions and subsequent warnings no doubt saved hundreds of lives.

But not all quakes have the same physical precursors as the Haicheng quakes. Earthquakes undoubtedly do have precursor signs, but they tend to be clearer in hindsight, and it is difficult to sort out the meaningful signs from the ones that lead to a dead end. In fact, after the success of predicting the Haicheng and Gushan-Pianling quakes, Chinese scientists proved this: no one predicted one of the deadliest quakes in China's history, which occurred on July 28, 1976. And that Moment Magnitude M_W 7.6 earthquake in Tangshan killed an estimated 240,000 people.

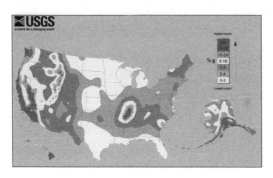

The earthquake hazard map of the United States. Darker spots indicate more potential for earthquakes. Courtesy of USGS.

In the meantime, other physical ways of forecasting a quake have surfaced. One controversial method was devised by Greek researchers from the University of Athens. Known as the VAN method, after the initials of the scientists who developed it (Panayotis Varotsos, Kessar Alexopoulos, and Konstantine Nomicos), it uses electrical and magnetic activity in the ground to predict an earthquake—right down to the location, time, and magnitude of some quakes. Though the major objection to the method is that most researchers have had a difficult time reproducing the team's results, there was a recent success: according to scientists at the Earthquake Prediction Research Center at Tokai University in Japan, strange changes in the Earth's electrical and magnetic fields in the Izu Islands began at the end of March 2000. By the end of June, a series of earthquakes began, the strength of the signals increasing and reaching a peak just before a quake measuring magnitude 6.4 on the Richter scale struck on July 1, 2000. Both the magnetic and electric fields died down to normal levels after the quake.

> ## Predicting People
>
> Often a popular pastime for psychics and pseudo-scientists, earthquake predictions are claimed as "successes" when, in reality, the claim of psychic prediction is often given to fit the event–after the quake occurred. And then there's the size factor: many times a psychic will predict "a quake" along the San Andreas Fault for a certain day–and when a quake occurs, the predictor claims victory. In reality, the famous fault shakes almost daily–so a claim of a micro or minor quake would make the predictor "correct." In the world of earthquakes, it's the "big ones" that everyone wants to forecast–not background seismic activity. And there is no reasonable scientific method when it comes to forecasting a quake.
>
> Not every predictor leans on psychic powers. For instance, Jim Berkland, a retired Santa Clara, California, county geologist, has been in the earthquake prediction business since 1974. Most of Berkland's work is based on "prediction windows" of eight-day time periods; these are then based on syzygy (the conjunction or opposition of the Sun and Moon–or

Can Lions, Tigers, and Bears Predict?

One of the biggest debates in the scientific and nonscientific worlds is whether or not animals, and sundry other natural things, can predict the coming of an earthquake—whether near a plate boundary or in an intraplate zone. That's because, in various parts of the world, reports of dogs, cats, snakes, and horses "sensing" a quake before it arrived abound. As of this writing, and if you include humans, no living being has ever predicted a quake—at least, there is no instance that has been scientifically proven. We may want to believe it, but so far, no one has ever found an "earthquake" man, woman, child, or beast.

But there are people who believe there is something to animals

> when they are both on one side of the Earth or opposite each other, with the Earth in between, respectively), which occurs twice per month. Other factors enter into his prediction equation, including when the Moon is closer (perigee) or farther (apogee) from the Earth, tide heights at chosen stations, and several more parameters—including the number of lost pets from local newspapers. His predictions cover specific places—Mount Diablo and Los Angeles, California (covering a 140-mile radius around these two locations); the entire states of Washington and Oregon; and finally, the entire world. Is there any proof that Berkland can predict quakes? According to Roger Hunter, formerly at the USGS's National Earthquake and Information Center in Golden, Colorado, in a September/October 2006 issue of *Skeptical Inquirer*, there is no basis for Berkland's claims; statistically speaking, none of them show significant results when compared to chance. In other words, flipping a coin would be just as useful a tool for predicting quakes. There are others who do believe Berkland is on to something—but only time, and a few more predicted quakes, will tell.

and earthquake prediction. After all, if the electromagnetic signals from quakes (as mentioned above) are real, could animals sense that a shaking is on its way? Can they actually hear the quake coming, as their auditory capacities are well beyond the human range? Is it possible that animals can detect the P-waves generated by the rupture of a fault? Could they feel these waves as they hit first, before the S-waves, which cause the most damage?

According to those who believe the connection exists, more than about a hundred kinds of animals seem to be able to "predict" a quake. The most likely to foretell a quake are dogs, cats, chickens, horses, and other small animals—and on the larger end, elephants; less likely are goats, cows, and other large animals. No matter the species, there seems to be a story about how an animal behaved in a peculiar manner from up to a week to minutes prior to a quake: chickens stop laying eggs; bees leave the hive in a panic; cats meow plaintively; dogs bark or whine in agitation; mice appear dazed and allow themselves to be captured easily by hand; homing pigeons take longer to navigate to their destination; and even rats and weasels seem to be deserting the soon-to-be-shaken area.

But unfortunately, animals don't always react in the same way for each quake. In fact, it seems as if they only respond to certain shakings—and which kinds of quakes is still a matter of conjecture.

Even so, China seems to lead the pack when it comes to the animal–earthquake prediction connection. Just prior to the Haicheng earthquake in February 1975, people observed snakes coming to the surface rather than hibernating—and freezing to death in the days leading up to the quake. Even the Anshan Zoo in Liaoning Province is in the city's seismological network—reporting to the seismological bureau whenever they see the animals acting strangely. The massive 1976 Tangshan quake in China also had an abundance of animal–earthquake reports. One story related incidents of a thousand chickens in Baiguantuan all excitedly clucking and refusing to eat. Mice scurried around, looking for places to hide—and there were even reports of a goldfish jumping in and out of its bowl before the quake.

Japan has also had a long and special relationship with a certain animal and earthquake prediction—the catfish. The earliest known written record is a letter from Toyotomi Hideyoshi

(1536–1598), the unifier of Japan: around 1592, he wanted to build a castle in Kyoto and stated, "During the construction of Fushimi Castle, be sure to implement all catfish countermeasures" —apparently in reference to the thrashing around of the animal before a quake. But there is no real proof that these animals become agitated by quake shaking—though some people suggest that the fish sense minute changes in electrical currents generated underground before a quake.

Animal responses to quakes have also been reported by people in the United States—from the shakings that occur in California in modern times (like the Loma Prieta quake of 1989) to animals' responses in intraplate quakes. For example, during the major New Madrid quakes of 1811–1812, people noted that wild animals seemed to lose their fear of humans during those months; and horses reacted to the shaking by stopping or proceeding as if they were walking on ice.

GETTING A HANDLE ON THE MOVEMENT
Earthquakes on the Rise?

Consider yourself lucky; be glad you were not alive millions of years ago, as the Earth cooled. Then, earthquakes rocked the land; tsunamis rolled over beaches in frequent intervals; and volcanoes belched in response not only to the movement of the plates, but to hotspots that formed in pockets all over the Earth.

What about our modern world—is there a chance that earthquakes are on the rise? According to the USGS, we may seem to be having more quakes, but shakings of magnitude 7.0 or greater have remained fairly constant for the past century. Some even point to a decrease in the number of major quakes.

But there are a few reasons why it seems as if there is an increase in earthquakes.

The first is obvious: we are more hooked to information highways than ever before—from the television news, cell phones, and the Internet—making all the news right at our fingertips. Never before have we had access to every minute detail of a quake—either major or minor. There are even Web sites that will e-mail you when a magnitude 3.0 or larger quake occurs anywhere around the world (for those of you who are interested, check out chapter 12's resources section).

Second, there has been an increase in seismograph stations around the world, even as far away as above the Arctic Circle and at the South Pole. In 1931, there were a mere 350 stations around the world; today, there are thousands, most of them attached to high-speed data collection instrumentation, making it easy to pinpoint a quake within minutes.

More such stations are being integrated as you read this text: the Incorporated Research Institutions for Seismology (IRIS) is a university research consortium dedicated to exploring the Earth's interior through the collection and distribution of seismographic data. The IRIS Global Seismographic Network (GSN) is one of the four major components of the IRIS Consortium. The goal of the GSN is to deploy over 128 permanent seismic recording stations uniformly over the Earth's surface.

Something else has changed and increased, too—the human population. Around the year 2000, the world had more than 6 billion inhabitants, a doubling of the population in just the past 40 years. This population is not evenly distributed, either. Many live in areas where big quakes occur—most of them poor nations, with no money to build homes more resilient to quakes (or victims of shoddy and illegal construction in certain nations like Turkey). Building codes are almost nonexistent in these circumstances. Even in nations with enough money to know better—like the United States—few people are aware that building codes for earthquakes do not exist.

There is also another population concern when it comes to earthquakes: it is estimated that about 50 percent of the population will live along coastlines by the year 2025—and if that comes to pass, many of those people will be right in the path of future tsunamis.

And if these are quirks of nature, is there a way to protect ourselves, or to plan, or to predict when a major quake will strike such places? How can we be prepared, or even understand how to protect ourselves from such rarities?

Though we cannot predict earthquakes or other movements, there are some measures that can be taken almost anywhere we live. For example, in the case of earthquakes, certain groups have formed to inform the public and do research on the New Madrid fault system—the Center for Earthquake Research and Information at the University of Memphis is one. They not only watch the worldwide quakes, they report on quakes in the local area (such as the 3.9 and 3.1 magnitude earthquakes [a quake and aftershock] in Warm Springs, Arkansas, on October 21, 1999). Other studies are revealing a better understanding of the local geology, such as the recent study at the University of Colorado in which scientists believe they found a blind thrust fault—largely unknown until 1994, when the Northridge earthquake demolished parts of the Los Angeles area.

> **Can We Stop the Shaking?**
>
> Even as you read this, earthquakes are shaking the Earth. They may not be newsworthy, but they are still going on, evidence of our vulnerability to Nature's whims. And even as I write this, in a list of daily earthquake occurrences in the United States and those over magnitude 4.0 around the world, a quake occurred at 6:29 A.M. It was a magnitude 4.6 quake in the Mariana Islands region, site of the deepest trench known in the world, and a place where the Pacific tectonic plate is slipping under the Indo-Asian plate. The 137-mile- (220-kilometer-) deep quake was just a blip on the map. And that's the kind of quake we expect from our restless Earth: a plate-to-plate interaction creating earthquakes. There were other quakes this day, too—many in Alaska, some in California, and one in Nevada, all measuring no more than magnitude 3.7—scarcely a shiver. All explainable, all understandable.
>
> But what about the one on the island of Hawaii, shaking a mild magnitude 3.0, or the quake listed at 1:51 P.M. local time—a magnitude 3.3 in Tennessee, the epicenter at a depth of 3 miles (7.4 kilometers)? Or the shaking a few weeks later in Hawaii measuring 6.2—enough to shake apart some buildings

THE HUMAN FACTORS
Can the Enemy Be Us?

Humans are notorious for fussing with the planet. And even though we have yet to cut into more than about a few miles of the Earth's crust, people still often wonder if humans can be the cause of certain types of quakes. This possible human inducement has its roots in such locations as the United States, Japan, and Canada, in which certain industries have caused some "earthquakes"—not major ones, but big enough to cause some concern.

But how can humans create a quake?

and disrupt the islands for a few days? Yes, amid the usual quake-prone areas are the unusual, or so it seems.

Will humans ever be able to stop an earthquake along a plate boundary or in the middle of a crustal plate? Face it: we can't. Not only do we lack much of the knowledge as to why certain quakes occur, but we can't forecast them. Quakes along plate boundaries are a bit "easier" to understand, though they can't as yet be forecast. And quakes that occur out of nowhere or in the middle of a crustal plate seem to have a mind of their own.

What we can do is significantly mitigate the effects from a quake. Scientists continue to identify earthquake fault zones, uncover unconsolidated sediment likely to amplify earthquake waves, and determine unstable land that would be prone to sliding or liquefying during a strong shaking. On the physical structures side, engineers can design and build safer homes and offices; and government, university programs, and other concerned organizations can conduct reality tests on those designs—and invent a few better structural designs of their own.

What's the most important item on the "earthquake to-do list"? To educate the public so they can prepare and respond to moderate to major quakes.

Unlike the James Bond movie *A View to a Kill*, there are no bombs known that can rock a region. The movie's premise had the villain, played wonderfully by Christopher Walken as microchip maker (and really a KGB agent—this was filmed before the collapse of the Berlin wall) Max Zorin, who wanted to move both the Hayward and San Andreas faults in California in a "double earthquake." The movement would not only shake the inhabitants, but flood Silicon Valley, eliminating his competition. But there are no mines, or oil or gas wells in that area of the San Andreas and Hayward faults; and if there were, you could not dynamite the mines and cause water to "lubricate" the faults, inducing a quake—faults are already wet enough from groundwater. While we're on the subject of dynamite, it's puny in explosive power; so far, no explosion wrought by human hands can destroy an entire landscape, create an earthquake, and plunge the world into chaos.

Finally, destroying Silicon Valley wouldn't even make a dent in the microchip production department.

If dynamite and bombs won't do it, and we can't dig down that deep into the crust, how can anyone say humans can be responsible for quakes? There is one way: pumping liquids into what are called injection wells. The biggest precipitators are the injection of fluids into deep wells, mainly for waste disposal; still other fluids are pumped into oil regions in order to extract petroleum products.

That being said, the resulting quakes have not been earth-shattering. But there have been incidents of decent quakes from injection wells. One of the most famous occurred at the Rocky Mountain Arsenal northeast of Denver, Colorado—a well-known repository for noxious chemicals used in the development of chemical weapons. In 1961, a 12,000-foot (3,658-meter) disposal well was dug to deposit many of the chemicals used at the arsenal ("waste fluids," according to the U.S. Army); by 1962, the injections commenced. And by the end of 1962, about 190 small to moderate earthquakes had occurred. These tremors didn't stop even after the injections had ceased in 1966. But a year later, something moved, shifted, and settled, creating an earthquake measuring magnitude 5.5—the largest earthquake recorded in the area, and a shaking that was followed by additional aftershocks. The solution was only to become apparent by 1968, when the army slowly began removing fluid from the well in an effort to reduce the shaking. It worked: since that time, no earthquakes have been felt.

Though Colorado has only one major fault running through it—the Sange de Cristo—the state also has the distinction of having the most human-induced earthquakes in the world. Rangeley Field in northwestern Colorado once experienced quakes that apparently started from water flooding the field. To prove this, in the late 1990s, the USGS conducted a series of tests on abandoned wells. When they stopped injecting water into the wells, the earthquakes dropped from 50 to 1 or 2 a day; when they began injections again, the earthquakes reached about 50 a day as the pressure within the wells built up. The USGS no longer uses the site—and learned valuable lessons from the injection wells.

Not that others haven't learned: other shaky grounds due to injection wells are found in Paradox Valley. At this site, the U.S. Bureau of Reclamation was trying to stop salt from entering the Dolores River and eventually reaching the Colorado River. To stop this flow of the salt within the groundwater, the BOR drilled a series of wells all along the Dolores to stop the groundwater from flowing into the river—injecting the rejected water into a 14,000-foot (4,267-meter) well. The quakes started almost immediately. The program recorded about 4,000 quakes, but no one worried, as few were felt on the surface. Unbeknownst to the scientists, the earthquakes were building up—and by June 2000, a magnitude 4.5 quake occurred. The solution was not to stop the injections, but to slow them down: now the BOR injects into the well every other month. There are still quakes, but they are minor.

Real Life Imitating the Movies

Though it's true that James Bond and friends couldn't slick down the sides of the San Andreas Fault and cause chaos in the world, that doesn't mean someone hasn't thought about it in real life. Yes, back in the 1980s, scientists wanted to know if it would be possible to control the slippage of a fault merely by controlling the fluid pressure surrounding the rock. They decided to experiment with pieces of Berea sandstone from Kentucky, jokingly called the "lab rat of rock mechanics" because the samples are fairly uniform. Trying to simulate a fault (albeit on a much smaller scale), they took two samples, put them under pressure, and then injected water. When the samples cracked, the "fault" slipped more dramatically than expected—even more than if they had just increased the pressure on the dry rock. But no matter how much they tried, there was no way to control the fluid pressure in order to control the fault slippage. Again, as with most powerful things in nature, humans don't have a clue how to control the slipping of a fault.

CHAPTER 11

What Can We Do?

> *Ten years ago only astrologers, mystics and religious zealots were concerned with earthquake prediction. Today some of the most respected scientists in seismology are actively working on this problem. Increased knowledge of the earthquake mechanism has encouraged seismologists to believe that earthquakes are preceded by events that signal the coming of an earthquake within hours or days or years. The challenge comes in learning to recognize them.*
>
> —*Frank Press and Raymond Siever in their 1974 textbook,* Earth

WHAT'S THE RISK?
What We Fear the Most

Scientists from all over the world have conducted studies of earthquake-prone regions. No one can deny the fact that certain areas are more vulnerable to quakes than others. There is Tokyo's vulnerability to quakes, as the area is at the junction of three tectonic plates, making it a prime candidate for shaking. One of the worst earthquakes of the twentieth century killed more than a half million people: the Tangshan earthquake of 1976, which occurred on an earthquake belt in China. There are other quake-prone areas that are ripe for a quake, all lying right across major faults and plate boundaries, including Tehran, Sumatra, and Istanbul. But even with the knowledge of quake-prone areas, scientists still cannot predict when a quake will occur, and especially not how much energy it will expend. Such knowledge gaps in quake zones make it even more difficult to predict and understand quakes in the wrong places.

Earthquake Statistics
Unlike the measured rising and setting of the Sun and Moon, there is no firm timetable when it comes to earthquake events. Scientists have had to rely on recorded events (written by humans who often interpreted information based on their own agenda) and, if lucky, the interpretation of certain rock layers that reveal a catastrophic event. In more modern times, data from a growing yet still small number of seismographs and Global Positioning Satellite devices have helped record quake data. And though the amount of earthquake data available would probably reach from here to the Moon many times over, scientists can only rely on statistical data to interpret when a quake may occur in certain areas.

There is one fortunate fact we do know: larger earthquakes occur less frequently than smaller earthquakes. This relationship is actually exponential—for example, there will be 10 times as many quakes larger than magnitude 4.0 in a certain period than those larger than magnitude 5.0.

The path to the frequency of earthquakes leads again to statistics. Scientists at a country's geological surveys are usually responsible for determining when events may occur—and even may not occur. Similar to determining the possible flooding in an area—in which they mention statistically based 100- or 500-year floods—scientists work to determine the potential for quakes.

And as we all know, statistics can lie—or at least be very wrong. The frequency of such events can be based not only on historical records, but on physical evidence. Both of these factors are not only difficult to find, but also, often, hard to interpret.

> ### Comparing Quakes
>
> Scientists now believe the Sumatra quake was one of the largest recorded on Earth, but it may not top the Chilean quake of May 22, 1960. Depending on the study, many put the Chilean quake at an estimated magnitude of 9.5; the Sumatra quake is estimated at about Moment Magnitude M_W 9.3. But in other ways, the Sumatran quake beats out Chile: the December 2004 quake not only tore apart about 186 more miles (300 more kilometers) of its fault—808 miles (1,300 kilometers) to Chile's 621 miles (1,000 kilometers)—but its resulting tsunami killed more people because of the region's heavily populated coastlines.

Risk Taking
Determining the risk of earthquakes in the United States—and other unlikely quake places around the world—is a relatively young field. The major reason is the infrequency of the occurrences; unlike cities and countries along plate borders, in the lower 48 states, only California is along a major fracture zone. In fact, estimates are that only 5 to 10 percent of earthquakes occur away from a plate boundary, making them rare, but not impossible. Such gaps in data are a blessing in one way (it means there were no big quakes), but a scientific deterrent when it comes to statistics. On top of everything, the United States has only been in

existence for just over 200 years—and decent recorded data about quakes have only been gathered in the past century.

Another problem is the amount of equipment available to track the quakes; seismographs are increasing, but there are still gaps, and the mapping of subsurface faults is still expensive.

Probably the biggest reason why we know little about earthquakes in the wrong places in the United States is something we can't control—the terrain. Seismic waves tend to travel for long distances in the States, the result of relatively brittle and mostly flat-lying layers of sedimentary rock. This material carries seismic waves through thousands of miles of rock, even if there is merely a moderate earthquake. In fact, because of the underlying rock layers, damaging ground motion can occur in an area about 10 times the size of a comparable quake in California.

But there is one fact that scares scientists: 39 of our 50 states are vulnerable to earthquakes—and that doesn't include just the usual suspects near plate boundaries or near the hotspots of California, Hawaii, and Alaska. For example, one study conducted in 1990 showed that when differences in seismic wave attenuation are taken into account, the central and eastern United States have about two-thirds the likelihood of California to produce an earthquake with comparable damage area and social (and economic) impact within the next 30 years; within that time, it is estimated that there is a 40 to 60 percent chance that the quake will be a magnitude 6.0—enough to cause major damage, especially in the larger cities.

And it won't be anything like the quakes that hit the New Madrid or Charleston areas. It will be much worse. The population of the central and eastern United States has grown considerably—and that's an understatement—since the late nineteenth century. City populations have reached hundreds of thousands, even millions.

Protecting Ports

The United States has a vested interest in keeping its ports working—these international gateways to trade and commerce have always been important to our economy. But what would happen if there was a quake in the wrong place?

Researchers know that if a quake the magnitude of the 1906 San Francisco quake were to strike today along the San Andreas Fault (or other offshoot faults), the damage to such places as the Port of Oakland would be devastating. The U.S. economy would definitely suffer as the port was put back, piece by piece, into some semblance of order. Historians cite the magnitude 6.9 quake in Kobe, Japan, in 1995, a shaking that caused extensive damage to both the city and its port. At the time of the quake, the port was the sixth largest in the world; by 2003, it was listed as the thirty-second largest, and will probably never recover to its pre-earthquake condition.

Why was there so much damage? One of the major reasons was liquefaction: most of the Kobe wharves—like most around the world—are on loose soils, usually deposited as fill when the port was built. This unstable ground is prone to liquefaction when it is shaken, causing it to lose its strength and collapse large structures such as wharves, cranes, and warehouses.

What would happen if a violent quake occurred in eastern port cities in the United States, such as in Charleston, South Carolina, or Savannah, Georgia? Like their western counterparts, many eastern ports are built on loose gravel or ground, making a major earthquake damaging.

There was also a lesson about damage to coastal ports from Hurricane Katrina, the 2005 hurricane that wiped out numerous U.S. cities along the Gulf of Mexico and caused billions of dollars in damage from winds and rains. Officials learned that even if the physical damage to the port is minimal, there may still be repercussions from a major natural disaster—whether it is a quake or a hurricane: if there is major damage to houses and infrastructures, the labor force that keeps the port open and functioning will be greatly affected. In other words, the workforce to do the job will no longer exist.

It's All in the Details, Details

For years, seismic stations around the world have helped calculate the paths of, and accurately locate, earthquakes. There's enough data not only to fill all the rooms in the Pentagon, but also to create three-dimensional models of the mantle's structure like a CAT scan of the Earth. But there are gaps yet to be filled. For instance, the mantle below the Southern Hemisphere has yet to be deduced in any quantitative (or qualitative) way; in particular, scientists believe that the large-scale mantle upwellings called plumes are located beneath the larger oceans in that hemisphere. Now comes the potential of drifting seismic networks to the rescue.

These networks are seismic readings taken while drifting in the

---

Location Is Everything

Probably the one major device that has changed the field of seismology is the Global Positioning System (GPS), a group of satellites that allows an observer on the ground holding a device to calculate their location and height above sea level. With such devices in continuous operating mode, ground measurements can trace spots experiencing deformation over time scales measured in seconds. Add to this a satellite-based instrument called InSAR (interferometric synthetic aperture radar), and keeping track of the Earth's deformation—and often consequential quakes—just became easier.

InSAR has recently come into the forefront of instruments that measure changes in the Earth's shape—and so provides insight into several processes, such as subsurface movement of magma and the changes in strain before, during, and after an earthquake. With this data, scientists have made high-resolution maps of the deformation around an earthquake event, noting the patterns of the shape-shifting Earth weeks, or even years, after a quake occurs.

They've also gathered data for how magma moves near supposed dormant oceans. One such instrument is the MERMAID (an acronym for "mobile earthquake recorder in marine areas by independent divers"), which scientists hope will fill in the many earthquake monitoring gaps. If the MERMAIDs do the trick, hundreds of these instruments will drift submerged about 700 meters below the surface of the ocean, picking up relatively faint earthquakes, including the incoming P-waves, with arrival times measured in milliseconds. The data from these sensitive platforms will lead to images of the unmapped portions of the mantle, answering such questions as, does the mantle move as a whole or is it layered? How do mantle plumes work and move, and do they truly create hotspots? Or are there any cool spots in the mantle—any areas that differ greatly from our preconceived theories?

---

volcanoes, as instruments to measure such changes may not be possible to set up in remote or dangerous areas. InSAR also means speed in collecting important earthquake data: one can quickly survey remote sites with areas measuring thousands of square miles—and at the same time, gather images with resolutions of about 66 feet (20 meters). For example, in one study over just a few weeks, scientists assessed the deformation of more than 900 volcanoes in the central Andes. Similar ground-based measurements would have taken years—or even decades—to achieve. It's also a matter of money, as InSAR can take such readings of erupting volcanoes without the loss of expensive instruments; and there is also the matter of human lives, as such distant views eliminate the need to jeopardize scientists taking measurements in the field.

Though InSAR instruments are available today, there are not enough to cover the entire globe. Scientists envision constellations of InSAR satellites in a variety of orbits, including some that could stay right above strategic quake and deformation areas (in what's called a geosynchronous orbit), allowing nearly real-time images to help respond to natural hazards such as volcanic eruptions and earthquakes.

CHANGING OUR MINDS
Changing Times

The statistics on earthquakes are quite shaking: there are about 10,000 to 20,000 decent earthquakes registered by seismologists each year. Of those, an average of 17 cause casualties; and of those, only 1 or 2 cause more than 500 deaths. But though these statistics seem reassuring, they hide a significant fact: humans repeat the same process over and over in earthquake-prone and non-earthquake-prone areas—populations increase close to plate borders, and seem to build the same flimsy structures over and over. In non-earthquake-prone areas, the thought of quakes is often laughed at—even scorned when it comes to changing codes in new-home building. It's as if we are all 20 years old again, thinking nothing can happen to us. But it does—and, as always, when we least expect it.

Deaths and injuries from earthquakes come mostly from the obvious—the collapse of buildings and structures, and even the partial collapses that eventually give in to gravity. This thesis has been tested over and over in moderate to major quakes, and especially in third-world countries, where structures are less than earthquake-proof. Scientists have long studied the shaking, documenting and gathering information about how structures respond—something that will probably go on forever thanks to the irregular continuation of devastating earthquakes. "They happened before; they will happen again," is a seismologist's mantra.

There is no way to prevent an earthquake, but we can learn to cope with them. The information collected from each quake adds to the knowledge of just how quakes work, shake structures, and even tear apart underground infrastructures. It helps us to better understand what goes on inside a shaken building, how falling objects move, and even how buildings flex depending on the seismic wave. Instruments continue to be distributed not only in earthquake-prone regions, but in those places where quakes are rare. Only through earthquake data can scientists, engineers, builders, and even handymen who construct their own homes integrate better designs, plans, and building codes for earthquake-resistant structures. And with all this, in turn, researchers, government officials, and the rest of us hope to one day considerably reduce the loss of life and property from earthquakes.

Though the recording of quakes began in around 1880, the idea of installing instruments in buildings to record the shaking took until the 1940s. By the mid-twentieth century, more such devices were installed. By the early 1970s, the installations paid off—the San Fernando, California, quake of 1971 measured 6.6 magnitude on the Richter scale, ranking it as moderate to large, not great—but it shook a wide, heavily populated area, leaving 65 people dead and half a billion dollars in destruction in its wake. The quake was also a gold mine to scientists, revealing a wealth of information about how a building responds to a quake. And by the time of the Imperial Valley, California, earthquake on October 15, 1979, technology had caught up even more. The magnitude 6.6 quake produced solid data of how buildings and structures twist and turn during a major earthquake.

Today there are instruments installed in all sorts of places around earthquake zones—most, of course, around areas close to plate boundaries—in hospitals, and around bridges, dams, aqueducts, and other structures. According to the USGS, such devices are found throughout the earthquake-prone areas of the United States, including Illinois, South Carolina, New York, Tennessee, Idaho, California, Washington, Alaska, and Hawaii. In California, the California Division of Mines and Geology (CDMG) and the USGS operate instruments; the USGS operates the other instruments located in the seismically active regions of the nation.

> ## From Hurricanes to California's Big One
>
> Hurricane Katrina struck the Louisiana-Mississippi coasts in the summer of 2005, creating the most devastating natural disaster of the new century. The breaching of New Orleans levees flooded the below-sea level city. The coastline community of Biloxi, Mississippi, was virtually wiped off the map. Millions of families were affected—and though the cleanup will eventually come to an end, the scars on the land and people will last a lifetime or more.
>
> And geologists know that there eventually will be a "big one" in California—one that will mean mind-boggling devastation. It is this thought—and the lessons learned from Katrina—that has sparked scientists to work on how people can avoid the brutal effects of a future earthquake catastrophe.
>
> According to David Schwartz, of the USGS in California and the head of the USGS and San Francisco Bay Area Earthquake Hazards Project, there is a way for the city to prepare. According to Schwartz, scientists can't predict earthquakes in the short term, but they are better at forecasting long-range events. For example, to this end, more than 100 geologists, seismologists, geophysicists, and mathematicians put out a report in May 2003 citing the likelihood of another large San Francisco earthquake. The results indicated that there was a 62 percent chance that the region would experience a large-scale quake within 30 years.
>
> Thanks to the report, and the consequences of Katrina, the people and the local government of San Francisco have been listening, shoring up the city's infrastructure and making plans for disaster recovery for when the big one hits. The hope is that the process will help, but many scientists have another concern—if there is enough time before a big quake strikes.

Better Building
One of the first references to building codes comes from as early as 1750 BC, when the Babylonian Empire's Code of Hammurabi read, "If a builder has built a house for a man and his work is not strong, and if the house he has built falls in and kills the householder, that builder shall be slain." Of course, today's building rules and regulations are not *that* strict (though sometimes you wonder if there would be as many claims of shoddy workmanship if we had such rules).

California shows the best progression of building codes for earthquakes: at the time of the 1906 San Francisco earthquake, many municipalities had building codes, but no one even considered the possibility of seismic effects. Not surprisingly, the 1906 earthquake sparked the discussion to improve building design and to incorporate those improvements in regulatory building codes. Professional organizations, particularly the Seismological Society of America (formed in 1906), and later, the Structural Engineers Association of California, were instrumental in pushing the code provisions for earthquake-resistant construction.

The reasons are obvious: the coast of California and certain inland spots along the Sierra Nevada are notorious for quakes, with the Pacific tectonic plate moving almost 2 inches (4.6 centimeters) per year. This stress and strain has to be compensated for if people are to live in such a region. As the saying in seismic engineering studies goes, "Earthquakes don't kill people; buildings kill people." Thus the push to mitigate as much building collapse as possible during any quake.

By the end of the twentieth century, building codes were broken into two parts—the old and the new. Older buildings were modified as completely as possible to make them more earthquake resistant, whereas new structures were designed to be earthquake resistant from scratch to ensure the least amount of damage during a moderate to major quake.

So why does someone in Missouri care about what happens to a building during a quake in San Francisco? Why should they be concerned with the rupturing of a fault in Japan, or the trembling of the ground in Turkey? Because one day, it may happen to their state, to their home, office, schools, churches, and synagogue. Because earthquakes do happen in the wrong places, and you can learn from what others have experienced—especially when it comes to building codes.

What happens to a building during a quake? According to J. H. Rainer and T. D. Northwood in their paper "Earthquakes and Buildings in Canada" (written through the National Research Council Canada), during an earthquake, the ground moves in a random manner, in both horizontal and vertical directions. "Buildings are inherently strong in the vertical direction since they are designed to resist the vertical loads due to gravity, but they are relatively weak in the horizontal direction. Thus the horizontal forces and deformations assume a dominant role in earthquake engineering for buildings. There is an obvious similarity to the horizontal forces imposed by wind, and indeed protection against wind may sometimes constitute a large portion of the protection needed against earthquakes. An essential difference exists, however: buildings are usually designed to deform elastically up to the design wind load, whereas some inelastic deformation is expected at the design earthquake load; attention must therefore be given to the ductile performance of the structure."

In other words, designing earthquake-resistant buildings is no easy exercise. Not only do the design engineers know how structures move, but they also know about the stresses caused by the shaking. Helpful instrumentation should not only be in the structures, but also on the nearby ground, to measure the moving earth and the shaking building. And every time a strong earthquake strikes, the new information gathered enables engineers to refine and improve the previous structural designs and building codes.

Modern criteria for seismic design and construction have been included in what is called the Uniform Building Code since 1973. The 1997 edition has the most up-to-date requirements, including greater strength for essential structures (such as hospitals) and for sites on soft, loose soils where shaking intensity is increased. The code sets minimum requirements that ensure life safety, but also allow for some earthquake damage and loss of function.

And those homeowners who hope to have less potential damage—and to be able to live in the building after a severe quake—should insist on higher standards for design, construction, and inspection for not only their current home, but also for a home they may have built in the future. You have to be the one who decides what level of damage is acceptable—something that a civil or structural engineer, and architect (yes, there are earthquake architects out there) should know.

New Madrid and Building

What about those people in other, less active, quake regions, such as the New Madrid zone? Should these areas also revamp their building codes just in case a major quake occurs? One current discussion concerns the latest building code, IBC 2000—a national code developed under the direction of FEMA. This code provides not only for buildings to withstand a possible major quake, but also for seismic safety—making the interior environment of the home safer from quakes, too.

But FEMA estimates that buildings in cities in the zone, such as Memphis, are only 5 to 10 times less likely to be damaged than those in San Francisco or Los Angeles. This has led some researchers to say that FEMA is making unrealistic specific building codes around the New Madrid zone and that their data about the damage from a quake are exaggerated. They cite such facts as frequency of quakes (New Madrid quakes occur about 30 to 100 times less frequently than quakes in Southern California); and the motion of the plate within the boundary zones (the Pacific plate moves about 1.8 inches [4.6 centimeters] per year versus the less than just over 0.5 inch [2 centimeters] around the New Madrid zone). Add to this the fact that a building in California will experience much more in, say, 50 years, while one in the New Madrid zone may shake once or twice in the same period.

They also cite the fact that because earthquakes of a given magnitude are about 10 times more frequent than those one magnitude unit larger, the shaking difference reduces the effect of the difference in quake rates by a factor of 10. The net effect of these differences depends on the recurrence rate of large quakes and the resulting ground motion—and currently, no one has a firm grip on how either of these factors works. Therefore, buildings in California are much more likely to be seriously damaged than those in the New Madrid zone.

On the other hand, if a major quake did occur in the New Madrid zone—and especially if the quakes were violent and continuous, as in the 1811–1812 quakes—they would be comparable to a California quake one magnitude larger. This is because the

Midwest layers of rock transmit seismic energy more efficiently. And if some scientists who study the New Madrid zone are correct, a big quake could happen within the next few decades. Do federal, state, and local officials truly want to tempt fate and not do anything about keeping buildings upright during a quake? It's a chance not many officials are willing to take.

Recognizing these problems, the USGS and other organizations have joined together to help greatly reduce loss of life and property in future temblors in the central states. According to the USGS, the following lists the past and latest implementations:

- In 1983, the states of Arkansas, Illinois, Indiana, Kentucky, Mississippi, Missouri, and Tennessee formed the Central United States Earthquake Consortium (CUSEC). CUSEC improves public earthquake awareness and education; coordinates multistate planning for earthquake preparedness, response, and recovery; and encourages research in earthquake hazard reduction.
- In 1990, the USGS, advised by private, academic, and government experts, issued a plan for intensified study of the New Madrid seismic zone. At the same time, the National Earthquake Hazards Reduction Program expanded efforts in the central United States.
- Earthquake education is now part of the curriculum in the schools of many CUSEC states. In Kentucky, the state legislature has mandated that earthquake education be taught in schools.
- Earthquake Awareness Weeks have been held in Arkansas and Kentucky for several years, and in Tennessee starting in 1995.

- Volunteer earthquake advisory councils or similar organizations have been formed in most CUSEC states.
- In 1993, with USGS support and collaboration, the CUSEC state geologists began a significant effort to map earthquake hazards. In 1995, they completed a regional soils map that can be used to locate areas likely to experience intense shaking in earthquakes.
- Most CUSEC states have adopted building codes containing modern earthquake design standards.
- Efforts to ensure the seismic safety of critical structures, such as dams, bridges, and highways, have accelerated. For example, in 1990, transportation agencies in Illinois, Kentucky, and Tennessee initiated programs to strengthen highway bridges that do not meet earthquake design standards.

In 1993, with USGS support and collaboration, the CUSEC state geologists began a significant effort to map earthquake hazards. In 1995, and later revised in 1999, they completed a regional soils map that can be used to locate areas likely to experience intense shaking in earthquakes because of loose or sandy soils. By 2003 and 2004, the CUSEC state geologists worked with the USGS to create an informative map, showing three centuries of significant earthquakes in the central United States. And even as this is being written, there are more CUSEC states and local jurisdictions adopting building codes containing the most up-to-date earthquake design standards.

Efforts to ensure the seismic safety of critical structures, such as dams, bridges, and highways, have also accelerated. For example, the USGS cites the transportation agencies in Illinois,

Kentucky, and Tennessee, which initiated programs to strengthen highway bridges that have not met earthquake design standards in the past. One includes the $175 million seismic retrofit of the Interstate 40 Bridge over the Mississippi River in Memphis, Tennessee—a necessary task that will allow the bridge to withstand large, damaging earthquakes.

> ### Mapping the Future
>
> One of the most important global earthquake-assessment collections in recent years was the Global Seismic Hazard Assessment Program (GSHAP), launched in 1992 and terminated in 1999. It was sponsored by the International Lithosphere Program (ILP) with support from the International Council of Scientific Unions (ICSU), and endorsed as a demonstration program in the framework of the United Nations' International Decade for Natural Disaster Reduction (UN/IDNDR). And now that the Internet has become so useful in spreading information, the USGS has jumped on the earthquake-mapping bandwagon and developed a few mapping projects of their own. For example, in 2005, the survey put out a new public Web site that graphically shows the probability of an earthquake shaking California in the next 24 hours. And knowing that the problem is not only in California, they have graphic maps of each state–and countries worldwide–that map the magnitude and location of quakes that have occurred in the last hour, week, and even six months (try it at http://earthquake.usgs.gov/eqcenter).

Quake-Proof Buildings Elsewhere

New England has had its fair share of micro to minor earthquakes in the past. And though seismologists believe the region carries a moderate risk for a quake, there have been some noticeable quakes in the past. The one that stands out in every New

Englander's mind is the Cape Ann quake of 1755—a magnitude 6.2 and called "the largest earthquake in recorded history in New England." Along a line from Boston to Montreal, it knocked down or damaged as many as 1,600 chimneys, collapsed the brick walls of many buildings, and toppled stone fences throughout the countryside. If such a quake were to occur today, it could kill hundreds of people and cause thousands of injuries—not to mention $6 billion in property damage.

Scientists still don't know if New England is due for a strong (intensity 6.0–6.9) or moderate (5.0–5.9) earthquake. It may sound like waffling, but scientists have tried to narrow down the answer—no, there probably won't be a strong quake, but yes, there could be a moderate quake. They know that earthquakes with a magnitude 6.0 or greater occur in New England on average once every 450 years—so the region is probably safe for 200 to 300 more years. But a quake measuring around magnitude 5.0 hits the region every 50 or 60 years. With the last one in 1940 near Ossipee, New Hampshire, some scientists believe there is a 19 to 28 percent likelihood of a moderate quake in New England by 2013—and a 41 to 56 percent likelihood by 2043.

Because of this possibility, many New England states (but not all) have been modifying their building codes to prepare for a possible shaking. For example, Massachusetts is one of the few states in the East whose building code contains a seismic provision. Instated in 1975, it has ensured that all buildings constructed since then can withstand a moderate-size earthquake. But this applies only to newly constructed buildings—and older buildings aren't subject to such stringent rules. For example, Boston and other areas in the state have many unreinforced buildings from the nineteenth and twentieth centuries that were not built to withstand a quake. And a number of structures in Boston are built upon large areas of landfill, which amplifies the effects of the tremors.

What about nearby Canada? Large earthquakes have occurred in areas adjacent to the Saint Lawrence River, on the West Coast, and in the northern regions, and can be expected again. In order to limit the risk to life and property, it is prudent to anticipate the consequences of a projected seismic event and to employ principles of seismic-resistant design; minimum requirements for this purpose are given in building codes such as the National Building Code of Canada. Generally the emphasis has been on protection of the buildings' occupants rather than on prevention of structural damage. If in specific instances greater seismic resistance seems appropriate—to guard against some special hazard—more stringent measures can and should be adopted.

Getting Rid of Hazards
There is no way to stop an earthquake—either along plate boundaries or in the wrong place. But there are ways to try to mitigate the damage and loss of life. One is the above-mentioned CUSEC, founded in 1983, with funding support and in partnership with FEMA. Another organization is the National Earthquake Hazards Reduction Program (NEHRP). Under the NEHRP, the USGS is mandated to monitor quakes and provide earthquake warnings and notifications. It's the only

government agency with this nationwide task—from the continually shaking California to the potentially hazardous New Madrid and Charleston regions. This group has a monitoring system that provides warnings, assesses seismic hazards, records quake activity, and provides the information that most places need to design building codes for new construction and to retrofit existing structures. They also provide timely information to emergency response teams in case of an earthquake, sending out data on the distribution and severity of the shaking. Such data are further used in the design and construction of safer, more earthquake-resistant future buildings and structures.

There are also some initiatives that have started to improve earthquake monitoring and reporting while trying to minimize the economic impact of such events. One recent effort is the Advanced National Seismic System (ANSS), in which about 500 new earthquake monitoring instruments have been set up in potentially vulnerable urban areas, including San Francisco, Seattle, Salt Lake City, Anchorage, Reno, and Memphis. The goal is to establish a network of 7,000 such devices around the country—especially in places like Memphis that aren't on the "earthquake-prone" list. Once in place, it will provide emergency response personnel with real-time (that is, within about 5 to 10 minutes of the initial earthquake event) information on the intensity and location of the shaking. This can be used to guide emergency response efforts, especially for rescue and recovery teams. The data collected can also be used by engineers and builders all over the United States, equipping them with guidance information to build more earthquake-resistant structures and retrofit existing ones.

> ### The Importance of Being Insured
>
> How do quakes compare with other natural disasters? The chart* below shows that earthquakes are number one when it comes to catastrophes:
>
> *Size of event in terms of fatalities*
>
> | Earthquake | Flood | Volcano | Frequency of event |
> |---|---|---|---|
> | 1 in 100 years | 472,000 | 98,000 | 34,000 |
> | 1 in 500 years | 1,052,000 | 520,000 | 74,000 |
> | 1 in 1,000 years | 1,446,000 | 1,061,000 | 97,000 |
>
> * *Nature,* December 15, 2005

Why is such information so important? It comes down to money, as usual. This particular chart was written by risk-management specialists to help insurance companies understand the scope of these disasters—because eventually, whether quakes occur in the expected places or not, someone has to pay for the cleanup.

According to the USGS and the Alaska Sea Grant College Program, the most common type of earthquake insurance is normally added as an endorsement on a standard homeowner's insurance policy. They state: "Typically, there is a deductible of 5 to 10 percent of the value of the home. This means that for a home currently insured at $150,000, you would have to pay $7,500 to $15,000 on damages before the insurance company would pay anything. Separate deductibles may apply to the contents of the house and the structure. An important coverage is temporary living expense, which pays for motel and meals if you have to move out of your home. There is usually no deductible on this coverage. The yearly cost of residential earthquake insurance is normally about $1.50 to $2.00 per $1,000 of coverage on a conventional frame home. However the rate may rise to $6.80 to $12.80 per $1,000 of coverage on structures with brick or masonry veneer on the outside. Clearly, the insurance industry considers homes without brick or masonry to be better risks in an earthquake."

There are other things to be aware of when thinking about insurance and earthquakes. For example, most policies do not cover damage as the result of an earthquake—and even existing fire insurance often does not cover fires caused by earthquakes. This was seen in the 1989 Loma Prieta earthquake: it caused over $6 billion in damage, but insured property damage accounted for only 16 percent of this loss.

But every state is different. The best course of action, of course, is to talk to your insurance agent.

Don't Be Scared; Be Aware
"It only takes one."

This quote came from an official at FEMA—and it's dead-on accurate. It only takes one major earthquake in the wrong place to create massive destruction, especially for cities that decided long ago to ignore the warnings.

We all saw this happen with a different kind of natural disaster, when Hurricane Katrina hit the Gulf Coast in 2005. For years, the city of New Orleans was warned about the possible flooding if a category 5 hurricane—the highest and most dangerous—struck. It was a disaster waiting to happen, as the city lies so many feet below sea level. The hurricane devoured cities along the coast;

floodwaters rose from a few to tens of feet in the city; and more than 3,000 people lost their lives.

The tragedy was nothing new to those of us who study natural hazards and geology. I even remember New Orleans being discussed as the "classic case" of a disaster-waiting-to-happen in graduate school. While we were studying the effects of subsidence due to sediment buildup from the Mississippi River, we also knew, decades ago, that a major hurricane would literally sink the city—with or without levees and channels put in by the Army Corps of Engineers. A direct hit would be like filling a bowl with water and realizing the water could not escape.

Though there has been significant progress in earthquake research, monitoring, and reporting, the risk of earthquakes is still high in many regions—especially in third-world countries in which the population growth and lack of earthquake-resistant structures go hand in hand with disaster. The problem, as always, comes when humans try to coexist with nature. We have to realize that no matter how much we think we've "reined in" nature, hazards like earthquakes will always be with us.

How do we become more aware? As the population increases, an out-of-nowhere earth movement has, and will continue to have, devastating effects. But we do have one advantage: there has been an increase in studies of such earth movements—from earthquakes and volcanoes to sinkholes and asteroids. This includes an increase in technology watching the areas for any signs of movement—and more education, to make the public more aware. And there are also educational efforts out there—including the Missouri Earthquake Awareness Week, which takes place at the end of January and emphasizes the possibility of another major earthquake in the New Madrid area.

Education is essential—making the public aware that most natural disasters are not random acts, but are direct, and often predictable, consequences of improper land use. Why build those houses along the San Andreas Fault—or eliminate special building codes for those living near the New Madrid fault line? In the end, when a disaster occurs, it is the taxpayers who pay the enormous long-term costs for such inappropriate building and lack of responsibility in our structures and infrastructures. The up-front cost of education may seem high in the beginning, but it will pay in great dividends when a natural disaster strikes—in terms not only of dollars but of lives.

We need to realize that earthquakes in the right and wrong places are merely natural processes that have been going on since the beginning of our world. We cannot control them; we cannot make them go away. But we can become "street-smart" about the possibilities—using lands in constructive ways so as not to invite death and destruction. We know the risks, and as technology improves, we'll continue to understand even more of how and when the Earth moves in all the wrong places.

And when that happens—and it will, many, many times in the future—we'll know what to do.

CHAPTER 12

The Best Defenses Are Good Resources

With all the destruction they cause and the problems with prediction, earthquakes are still amazing events. It's as if the Earth is ringing, fluctuating, flinching—almost alive, trying to get comfortable in its travels through the galaxy and the universe. The Earth in motion has affected humans for centuries—and no doubt (to a point) will continue to do so for centuries more. One of the major reasons is the growth of our populations—and the necessity of shelter. Some areas may seem safe for all of us, but in reality, there are hidden problems that could potentially affect millions. They have before; they will again.

For those of us living in a major or minor earthquake zone, the answer to a possible shaking will be how well prepared we are to face the quake. Some of us will contribute by monitoring any earthquake we feel; others will help by knowing what to do—and helping others—if a moderate to major quake occurs. And in order to be prepared, you need a few good resources—from what to do during and after a quake to the best Web sites, books, and articles to consult.

GIVING AND GETTING HELP
Citizen Science: Can You Feel It Coming?

One of the best ways to keep track of earthquakes in the wrong places is people. Yes, it may be that increases in population will cause more problems in future quakes—more people equals more damage and death—but it will also be necessary for citizens to help scientists learn everything they can about quakes.

One of the first attempts at such "citizen science" was started by the USGS and several of the regional seismic networks in California. A while back, the public learned about quakes through the radio and television media; with today's extensive Internet connections, such information is right at a person's fingertips. For example, the USGS site allows immediate and daily access to all known earthquakes around the world. Following this site, you'll know if there has been even the most minor tremor in your own backyard—the magnitude, intensity, longitude, latitude, and often details about how the USGS interprets the earthquake.

But just knowing that a minor or major quake has occurred is not enough. Scientists want to know more about such shakings—especially in places in which earthquakes are less common. To accomplish this, the USGS site also has a section called "Did You Feel It?" that allows people to share their information about the effects of a quake in their area. Visitors to the site can enter their zip code and answer a list of simple questions, such as, "Did the earthquake wake you up?" and, "Did objects fall off the shelves?"

Scientists collect the answers to these questions and essentially build an Internet version of the Modified Mercalli scale, converting the responses to equivalent intensities for each region.

Sometimes within minutes of a quake, the data are interpreted, creating a map of intensities; it is further updated as more data are collected and more is known about the quake. These "Community Internet Intensity Maps" contribute greatly to determining the scope of an earthquake emergency, especially in regions lacking seismic instruments. They also contribute to the data sets of regions that rarely experience quakes. You can enter this world of citizen earthquake science by visiting http://earthquake.usgs.gov/eqcenter/dyfi.php.

Preparing for a Quake
Even if you don't live in an earthquake-prone area, that doesn't mean you will never experience an earthquake. And if you do, the quake will probably be just a rumbling at your feet. But there is always the potential to experience a moderate—or even major—quake. How can you prevent injury and protect your property? FEMA has a few suggestions:

- Bolt bookcases, china cabinets, and other tall furniture to wall studs. Brace or anchor high or top-heavy objects. (During a quake, these things can topple over, causing injury.)
- Secure items that might fall, such as televisions, bookcases, and computers. (During a quake, these objects can act like projectiles if they're small enough, such as laptop computers.)
- Install strong latches or bolts on cabinet and closet doors. (The quake could shake open doors, causing items to fly out.)
- Move large or heavy objects and fragile items to lower

shelves. (They are less apt to fall and injure someone during a quake.)
- Store breakable items such as bottled food, glass, and china in low, closed cabinets with latches. (Again, they are less apt to fall and injure someone during a quake.)
- Store weed killers, pesticides, and flammable products securely in closed cabinets with latches, and keep them on the bottom shelves.
- Hang heavy items, such as pictures and mirrors, away from beds, couches, or anywhere else people might be sitting. (A quake happens rapidly and "out of the blue"; if you're sitting below a heavy picture or mirror, you can be injured during a quake.)
- Brace overhead light fixtures. (They can fall during a quake.)
- Strap your water heater to wall studs. (You want to protect this tank—it may be your best, and only, source of drinkable water following an earthquake, so protect it from damage and leaks.)
- Bolt down any gas appliances. (A ruptured gas line can create a fire hazard; by bolting down the appliance, there is less chance that the line will rupture if the stove, etc., moves.) Along with this, put in flexible pipe fittings to avoid breakage during a quake.
- Repair any deep cracks in your ceilings or foundations. Ask an expert if there are any signs of structural defects. (A small hole or crack can become a large one during and after an earthquake.)
- Check to see if your house is bolted to the foundation; if so, it is less likely to be severely damaged during an

earthquake. (Homes not bolted down have been known to slip from their foundations during a quake, making them uninhabitable.)
- Check with a professional structural design engineer to see if your home or building is in good condition. Ask for strengthening tips for porches, front and back decks, sliding glass doors, canopies, carports, garage doors, etc.
- For those of you potentially in harm's way during even a moderate quake, you might want to look into earthquake insurance.

For more information about earthquakes in your region, call FEMA's toll-free help line (as of this writing; we apologize if this number has changed) at 1-800-621-3362 (FEMA).

During a Quake

FEMA has some specific rules for getting through an earthquake. Here are the highlights:

- If you are inside a building: Stay there and cover your head with your arms. Protect yourself from falling debris by positioning yourself in a strongly supported doorway, under a sturdy table, or against an inside wall within three seconds or less. If you cannot immediately find a table or doorway, cover your head with your arms and crouch in an inside corner of a room. Stay there until the shaking stops.
- Center of room: However, when a house falls over during an earthquake, doorways are frequently more impacted than other areas of the house. In severe earthquakes, the center of the room is usually a safer place

> ### Drop, Cover, and Hold
>
> The Boy and Girl Scouts had it right: be prepared. One of the ways FEMA and other safety organizations grab your attention, and help you to remember things in a panic situation, is with a cute phrase or motto. For a flash flood situation with water rising over a road, the NOAA Weather Service suggests, "Turn Around, Don't Drown." For surviving an earthquake, FEMA's motto is, "Drop, Cover, and Hold On."
>
> And it makes perfect sense: DROP down on the floor. Take COVER under a sturdy desk, table, or other piece of furniture. (If that is not possible, seek cover against an interior wall and protect your head and neck with your arms; avoid danger spots near windows, hanging objects, mirrors, or tall furniture.) If you take cover under a sturdy piece of furniture, HOLD ON to it and be prepared to move with it. Hold the position until the ground stops shaking and it is safe to move again.
>
> And as the National Division of Homeland Security and Emergency Management adds: "No matter where you are, know how to protect yourself and your family during an earthquake. Practice taking cover as if there were an earthquake and learn the safest places in your home and work. Practice getting out of your home and check to see if the planned exits are clear and if they can become blocked in an earthquake. Practice turning off your electricity and water. Know how to turn off the gas, but do not practice this step. In the event of an earthquake, once you turn off your gas, only your utility company should turn it back on for safety reasons."

than doorways. There are exceptions to this. People in unreinforced brick buildings, in a building with two or more stories, or in an apartment that may implode due to heavy weight should not consider the middle of the

floor as a safe place. Stay away from glass, windows, mirrors, outside doors and walls, and anything that can fall. Do not use the elevators.
- If you are outside: Go to an open area. Steer clear of power lines, posts, walls and other structures that may fall. Stay away from buildings with glass panes.
- On the road: When driving, decrease speed gradually and pull to the side of the road and stop. Avoid stopping near buildings, trees, and utility wires. Do not attempt to cross bridges or overpasses. Stay inside the vehicle and keep your seat belt fastened.
- Miscellaneous places: If you are on a mountain, move away from steep slopes. If you are along the shore, run toward higher ground. When trapped under debris: Do not light a match. Do not move about or kick up dust. Cover your mouth with a piece of cloth. Tap on a pipe or wall so rescuers can locate you. Use a whistle, if available. Shout only as a last resort, because you might inhale dangerous amounts of dust.

HELPFUL WEB SITES AND RESOURCES
Great Web Sites
One of the best ways to keep up with the latest earthquakes, ground movements, and sundry quake effects—especially in the wrong places—is via the Internet. Here are a few Web pages used in researching this book and some for those who want to know more about quakes that occur out of nowhere (please note: Web sites change periodically; I apologize for any inconvenience caused by changed addresses or sites):

General Earthquake Information

United States Geological Survey—the best geologic site around, with natural hazards information at your fingertips; you can even sign up and have the USGS e-mail you when a magnitude 3.0 or larger earthquake occurs anywhere around the world.
www.usgs.gov

Electronic Encyclopedia of Earthquakes, containing a digital library of earthquake information, curricula, and resources for educators, students, and the general public.
http://sceccore.usc.edu/e3/index.php

Science Education Resource Center, a tremendous project through Carleton College in Minnesota that helps students and teachers understand the geology and physics behind many natural events—including earthquakes and volcanoes; a must-check-out site for animations of various geologic events, such as the movement of the tectonic plates over time. This is the address for their "Teach the Earth" section:
http://serc.carleton.edu/index.html

EarthScope is a group that combines geophysical measurements with data and observations from all disciplines of the Earth sciences to extend the knowledge of North America's structure and evolution—and it includes work on earthquakes, too.
www.earthscope.org

EarthScope again, just for fun. (This is with **UNAVCO,** a non-profit, membership-governed consortium funded through the

National Science Foundation and NASA. The UNAVCO is constructing the EarthScope Plate Boundary Observatory, a geodetic observatory designed to study the three-dimensional strain field resulting from deformation across the active boundary zone between the Pacific and North American plates in the western United States.)
www.unavco.org

NOAA Satellite and Information Service's National Geophysical Data Center (earthquakes, volcanoes, and tsunamis, too).
www.ngdc.noaa.gov/seg/hazard/hazards.shtml

Historic Earthquakes

United States (even listed by state) through the United States Geological Survey: http://earthquake.usgs.gov/regional/states

HEAT—Historic Earthquake Theories (Konrad Lorenz Institute, Austria)
www.univie.ac.at/Wissenschaftstheorie/heat/heat.htm

The Virtual Times: New Madrid Earthquake (with many other interesting links)
http://hsv.com/genlintr/newmadrd/

New Madrid Compendium (great historical record of New Madrid quakes)
www.ceri.memphis.edu/compendium/

Quakes in Central United States (including the New Madrid fault zone)

List of Recent Earthquakes in Central States:
http://folkworm.ceri.memphis.edu/recenteqs/Quakes/quakes.big.html

Cooperative Mid-America Madrid Seismic Network (out of Saint Louis University):
www.eas.slu.edu/Earthquake_Center/NM

Quakes in the Northeast and Southeast

Seismo-Watch—Northeast United States Earthquake News and Information (includes Canada, too)
www.seismo-watch.com/Regions/na/usa/neus/neus.html

New England Seismic Network (out of the Earth Resources Laboratory in Cambridge, Massachusetts)
www-eaps.mit.edu/erl/research/NESN.html

New Hampshire Department of Safety earthquake page
www.nhoem.state.nh.us/NaturalHazards/Earthquakes.shtm

Lamont-Doherty Cooperative Seismographic Network (Columbia University, New York)
www.ldeo.columbia.edu/LCSN

New York Earthquake information compiled by Dee Finney (2002; some good resources)
www.greatdreams.com/ny/newyork_quakes.htm

Maryland Seismic Network
www.mgs.md.gov/seismics/index.shtml

Natural Resources Canada (earthquake page)
http://ess.nrcan.gc.ca/index_e.php

Other Quakes around the World

Pacific Northwest Seismic Network
www.pnsn.org/welcome.html

Southern Ontario Seismic Network
www.gp.uwo.ca/welcome.html

Middle America Seismograph Consortium
http://midas.upr.clu.edu

European-Mediterranean Seismological Centre
www.emsc-csem.org/index.php?page=home

Institute of Seismology (an independent institute under the Senate of the University of Helsinki)
www.seismo.helsinki.fi/index_uk.html

Seismology Research Centre, Australia (latest earthquake news in Australia)
www.seis.com.au

Regional Center for Seismology for South America (CERESIS)
www.ceresis.org/new/es/index.html

Laboratorio de Geofísica, Universidad de Los Andes, Venezuela (geology lab)
http://lgula.ciens.ula.ve

International Charter "Space and Major Disasters" (information about the December 2004 earthquake in Sumatra)
www.disasterscharter.org/disasters/CALLID_077_e.html

Edinburgh Earth Observatory's World-Wide Earthquake Locator (excellent site)
http://tsunami.geo.ed.ac.uk/local-bin/quakes/mapscript/home.pl

Center for International Disaster Information (pick your natural hazard)
http://iys.cidi.org/disaster/

Meteoquake Research Centre (United Kingdom)
www.meteoquake.org/res1.html

Benfield Hazard Research Centre (United Kingdom)
www.benfieldhrc.org/

Advanced National Seismic System (worldwide)—Composite Earthquake Catalogue
www.ncedc.org/cnss/

Earthquake Maps on the Web
http://cires.colorado.edu/people/jones.craig/Web_EQs.html

Volcanoes, Tsunamis, and Landslides

Global Volcanism Program (USGS)
www.volcano.si.edu/reports/usgs/

United States Geological Survey's National Landslide Information Center
http://landslides.usgs.gov/nlic/

International Landslide Centre (United Kingdom)
www.landslidecentre.org/

International Tsunami Information Centre (hosted by NOAA from Hawaii)
http://ioc3.unesco.org/itic/

NOAA Center for Tsunami Research
http://nctr.pmel.noaa.gov/

Pacific Tsunami Museum (Hawaii)
www.tsunami.org/

Alaska Volcano Observatory
www.avo.alaska.edu/

Hawaiian Volcano Observatory (run by the USGS)
http://hvo.wr.usgs.gov/

Quake Prediction

SyzygyJob—Jim Berkland's page for earthquake prediction; he was also the main subject for the book *The Man Who Predicts Earthquakes* (Cal Orey, Sentient Publications, 2006)
www.syzygyjob.com/

Earthquake Prediction Research Center (an interesting site out of Japan)
www.eprc.eri.u-tokyo.ac.jp/EPRC_E.html

Quake Coping

Rensselaer Polytechnic Institute's study on wood-frame houses and quakes
www.newswise.com/p/articles/view/521386/

Earthquake Engineering Research Institute
www.eeri.org/

Southern California Earthquake Center
www.scec.org/

Earthquake safety preparedness information
www.quakeinfo.org/

National Information Service for Earthquake Engineering (from the University of California at Berkeley)
http://nisee.berkeley.edu/

Consortium of Universities for Research in Earthquake Engineering
www.curee.org

Best of the Books and Articles
There are plenty of books about natural hazards—including earthquakes, volcanic eruptions, landslides, and tsunamis. Just type in any of those keywords into the search engines at Amazon.com and Barnes&Noble.com. The results include not only books and journals, but DVDs from such groups as National Geographic and Discovery Channel.

In the meantime, here are a few of the books I enjoyed that pertain to earthquakes in the wrong places as well as offering general information about quakes, from how to help yourself in a quake to how to earthquake-proof your home:

Atkinson, William. *The Next New Madrid Earthquake: A Survival Guide for the Midwest.* Carbondale: Southern Illinois University Press, 1989.

Bagnall, Norma Hayes. *On Shaky Ground: The New Madrid Earthquakes of 1811–1812,* Columbia: University of Missouri Press, 1996.

Barnes-Svarney, P., and T. Svarney. *A Paranoid's Ultimate Survival Guide.* New York: Prometheus Press, 2002.

Cote, Richard N. *City of Heroes: The Great Charleston Earthquake of 1886.* New York: Corinthian Books, 2006.

Decker, Robert, and Barbara Decker. *Volcanoes.* New York: W. H. Freeman and Company, 1998.

Evernden, J. F., ed. *Abnormal Animal Behavior Prior to Earthquakes.* U.S. Dept. of Interior Geological Survey, Conference I. Convened under the auspices of the National Earthquake Hazards Reduction Program, USGS, Menlo Park, CA, September 23–24, 1976.

Feldman, Jay. *When the Mississippi Ran Backwards: Empire, Intrigue, Murder, and the New Madrid Earthquakes.* New York: Free Press, 2005.

FEMA. *Homebuilders' Guide to Earthquake-Resistant Design and Construction,* FEMA handout, 1998.

Hough, Susan, and Roger Bilham. *After the Earth Quakes: Elastic Rebound on an Urban Planet.* New York: Oxford University Press, 2005.

McKeown, F. A., and L. C. Pakiser, eds. *Investigations of the New Madrid, Missouri, Earthquake Region.* Washington: U.S. Government Printing Office, 1982.

Page, Jake, and Charles Officer. *The Big One: The Earthquake That Rocked Early America and Helped Create a Science.* Boston: Houghton Mifflin, 2004.

Penick, James L. *The New Madrid Earthquakes.* Columbia: University of Missouri Press, 1976.

Van Diver, Bradford B. *Roadside Geology of New York.* Missoula, MO: Mountain Press Publishing Company, 1985.

Winchester, Simon. *A Crack in the Edge of the World: America and the Great California Earthquake of 1906.* New York: HarperCollins, 2005.

Image Credits

p. 3	USGS, and cross section by Jose F. Vigil from *This Dynamic Planet*
p. 9	NOAA
p. 21	USGS
p. 23	Photo by Austin Post, Skamania County, Washington, courtesy of USGS.
p. 27	NOAA
p. 35	USGS
p. 41	NOAA
p. 50	Photo by Dane Golden/FEMA News photo.
p. 53	USGS Open-File Report 90-547 (ID Ellen, S.D., 3ct esd00003)
p. 87	Plate 2-A in the USGS. bulletin 494, 1912. (ID. Fuller, M.L., 375, fml00375)
p. 91	ID. Fuller, M.L. 353, fml00353
p. 107	FEMA News photo by Kevin Galvin
p. 109	ID Hillers, J.K. 14 hjk00014
p. 117	Plate 15, Annual Report 9 of the USGS 1887-88; ID Hillers, J.K. 14a hjk0014a

p. 121	USGS
p. 130	Ted Butts, March 1970
p. 154	USGS
p. 155	USGS
p. 157	From USGS Professional Paper 543-D
p. 158	U.S. Navy photo by Photographer's Mate 2nd Class Philip A. McDaniel.
p. 161	Courtesy of the Hawaiian Volcano Observatory through the USGS
p. 162	USGS
p. 163	USGS Open-File Report 90-547
p. 175	USGS
p. 176	September 1968, Dr. James P. McVey, NOAA Sea Grant Program
p. 188	NOAA
p. 209	USGS

Index

acceleration, 35
Adapazari (Turkey), 196
Adirondack Mountains (New York), 103, 120
Advanced National Seismic System, 242
Africa, 16, 30, 139–41
African plate, 10, 41, 122
aftershocks, 43–47, 194
Aguiar, Antone "Tony" Correa, *190*
Alaska
    Aleutian Islands, 8, 13, 27, 151, 167–69
    Andreanof Islands, 71
    Denali, 36, 38, 201
    earthquakes in, 36, *159*, 201, 202, 218
    and plate tectonics, 8
    Prince William Sound, 62, 71
    and the Ring of Fire, 27
    volcanoes in, 154–55
Albany (New York), 120
Aleutian Islands (Alaska), 8, 13, 27, 151, 167–69
Alpide belt, 27
Alps, 8, 10
Ambraseys, Nicholas, 53
Andes mountains, 12, 16, 141
Andreanof Islands (Alaska), 71
animals and earthquake prediction, 212–15
Antarctic plate, 141
Antarctica, 143–45
Appalachian Mountains, 10
Arabian plate, 10, 41
Arctic ridge, 18
Ardsley (New York), 4, 120
Aristotle, 64

Arkansas, 86, 91, 100, 217, 237–39
aseismic creep, 46
aseismic earthquakes, 203–4
aseismic fault slip, 46
Ashley River fault, 114
Asphaug, Erik, 183
asteroid strikes, 182–84
Atitlán volcano (Guatemala), 186
Atlantic Ocean, 170, 176
Attica (New York), 119
Au Sable Forks (New York), 4, 6
Australia, 43, 137–39
Australian plate, 8, 138, 172
avalanches, *159*, 188

ball lightning, 67
Barangay Guinsaugon slide (Philippines), 185–86
Basel (Switzerland), 136
Bath's law, 45
Berkland, Jim, 212–13
Big Bear (California), 195
Big Lake (Arkansas), 84
Bilham, Roger, 72
blind thrust fault, 30–31
body waves, 35–36, 61
bombs and earthquakes, 219
books and articles, 261–62
booming earthquakes, 68–69, 116
Boston (Massachusetts), 241
Brazil, 142
"breathing mode," 175
Brunhes Normal, 17

Bryah, Eliza, 87–89
building codes, 217, 233, 235–36, 240
buildings and structures
   and aftershocks, 44
   design and study of, 230–31, 233–36, 238
   earthquake preparedness, 249–53
   and loose soil, 40, 187
   and Modified Mercalli scale, 54
   and New Madrid earthquakes, 84
   and remotely triggered earthquakes, 197
   see also building codes
Burmese plate, 171–72

Cadoux (Australia), 139
Cairo (Missouri), 100
Calcutta (India), 72–73
California
   Big Bear, 195
   blind thrust faults, 30
   building codes, 233
   earthquakes in, 218, 231
   Fort Tejon, 38–39
   Hayward fault, 219
   Hector Mine, 155
   Imperial Valley, 231
   Lake Tahoe, 181
   Landers, 155, 195
   Loma Prieta, 35, 38–39, *55, 165,* 209, 244
   Long Valley caldera, 154
   Northridge, 30, 62, 101, 217
   Palo Alto, 3
   Parkfield, 195
   Paso Robles, *52*
   and plate tectonics, 8, 16
   San Andreas Fault, 13, 24, 219
   see also Los Angeles, San Francisco
Cameron's Line, 114, 122
Canada, 70, 119, 132–35, 179, 241
Cancani, Adolfo, 52
Cape Ann (Massachusetts), 130, 240
Caracas (Venezuela), 141–42
Caribbean plate, 176
Caruthersville (Missouri), 84
Cascade Range, 12, 151
causes of earthquakes, 6–19, 22, 34
   see also faults; plate tectonics
Center for Earthquake Research and Information, 217
Central America, 186

Central United States Earthquake Consortium, 237–38
Chagos Ridge, 161
Challenger Deep, 12
Charleston (Missouri), 91, 100
Charleston (South Carolina), 106, 110–14, *119, 123, 132*
Chikyu Hakken mission, 14
Chile, 43, 141, 154, 225
Chile rise, 20
China, 5, 202, 210, 214, 223
Cincinnati (Ohio), 92–94, 101
Cocos Plate, 20, 27, 40, 151
coda duration magnitude, 61
Colombia, 154
Colorado, 199–200, 220–21
comet strikes, 182–84
Community Internet Intensity Maps, 249
compressional waves, 35–36
continental drift theory, 14–17
   see also plate tectonics
continental ice sheets, 103, 123, 136–37
convection currents, 15–16, 18–19
convergent plate boundaries, 9–13
Cooter (Missouri), 84
Cordón Caulle volcano (Chile), 154
cores, inner and outer, 7, 19
Cornwall-Massena (New York), 118, 133
creep rates, 46
Crist, George Heinrich, 81, 87
crust, 7
crustal boundaries, 24
   see also plate boundaries
cryoseism, 75

Darwin, Charles, 193
de Rossi, Michele Stefano, 51
Deccan traps, 161
deformation of the earth's surface, 228–29
Delores River (Colorado), 221
Denali (Alaska), 36, 38, 201
dikes, 108–9
dip-slip fault, 26
divergent boundaries, 9, 13
Division of Seismology and Geothermal Studies (Canada), 133–35
Dow, Lorenzo, 88
Drake, Daniel, 92–94
drifting seismic networks, 228–29
"Drop, Cover, and Hold on," 252

Earthquake, The (Sutherland), 33
Earthquake Prediction Research Center (Japan), 211
earthquake storm, 193–98
"Earthquakes and Buildings in Canada" (Rainer and Northwood), 234
East African Rift (Alaska), 10, 30, 41
  see also Great Rift Valley
East Pacific rise, 18, 20, *29*, 44
education, 181–82, 219, 237, 245–62
Egen, P. N. G., 50
elastic rebound theory, 24
electromagnetic fields, 211
Eltanin impact, 184
Elysian Park, California, 30
English Channel, 136
epicenters, *11*, 35
  see also specific locations
Erzincan (Turkey), 196
Ethiopia, 41
Eurasian plate, 8, 12, 138
Europe, 136–37
European Macroseismic Scale, 58–59
European plate, 122, 151
European Space Agency, 137
Evansville (Illinois), 110
Explorer Ridge, *29*

faults, 13, 22–31, 43–44, 46, 222
  see also plate boundaries; specific faults
Federal Emergency Management Agency (FEMA), 6, 100, 235–36, 249–53
Florida, 3, 117–18
forecasting earthquakes
  see prediction
Forel, François-Alphonse, 51
foreshocks, 43–44
Fort Tejon (California), 38–39
fracture zone, 13
Franklin, Benjamin, 64–65
French Guiana, 142–43
frequency, 35
Freund, Friedemann T., 67
frost, 74–75
Fuller, Myron L., 90

Galápagos Islands, 20, 158
Galeras volcano (Colombia), 154
Ganymede, 78
Gauss Normal, 17
geysers, *177*

Gilbert Reverse, 17
glaciers, 103–4, 123
Global Positioning System, 203, 228
Global Seismic Hazard Assessment Program, 239
Global Seismographic Network, 216
Grand Banks (Canada), 133–34, 179
Grand Coulee Dam (Washington), 182
Great Rift Valley (Africa), 16, 139–40
  see also East African Rift (Africa)
ground displacement, 41–42
ground movements, 137
ground water, 200–202
Guatemala, 37, 186
Guinea, 140
Gulf of Aden (Indian Ocean), 41
Gulf of Mexico, 117–18
Gulf of Suez, 30
Gushan-Pianling (China), 210
Gutenberg, Beno, 45, 60
Gutenberg-Richter relation, 45

Haicheng (China), 210, 214
Halloween earthquake, 100
Hanks, Tom, 61
Hawaii (the Big Island), 158, 160
Hawaiian Islands
  earthquakes in, 154, 203–4, 218
  and hotspots, 151
  origins of, 158–61
  potential disasters, 190–91
  tsunamis, 71, 167–69, 171, *190*
  volcanoes in, 153–54, *163*, 190, 204
Hayward fault, 219
heaves, 74–75, 162
Hector Mine (California), 155
Heng, Zhang, 49
Hess, Harry, 16
Hilo (Hawaii), 167–69, 171, *190*
Himalayas, 8–10, 12, 16, 27
Holmes, Arthur, 15
Hong Kong, 106
hotspots, 151, 155–64, *177*
Hough, Susan, 44, 194–95
humans, effects of, 218–22
  see also population growth
Hunter, Roger, 213
Hurricane Katrina, 128, 227, 232, 244–45
Hurricane Stan, 186
hydrothermal systems, 162
hypocenter, 35

Iapetus Ocean, 123
ice-age, 103–4, 123, 136–37
Iceland, 8, 151, 202
Illinois, 107–10, 238–39
Imperial Valley (California), 231
Incorporated Research Institutions for Seismology, 216
India, 59, 72–73, 172, 186
Indian Ocean, 170, 176
Indian plate, 8–9, 12, 138, 161, 171–72
Indiana, faults in, 107–10
Indo-Asian plate, 218
Indo Australia plate, 10
Indonesia
    Krakatau volcano, 147, 150, 173
    and plate tectonics, 8
    and seismic activity, 27
    Sumatra-Andaman earthquake, 41, 59, 155, *160*, 174–75
injection wells, 220–21
InSAR (interferometric synthetic aperture radar), 228–29
Institut de Physique du Globe de Paris, 175
insurance, earthquake, 243–44, 251
intensity scales, 50–59
interferometric synthetic aperture radar (InSAR), 228–29
Internet and earthquakes, 135, 216, 239, 248–49
    see also websites
interplate earthquakes, 9
intraplate earthquakes, 9, 106, 131
intraslab earthquakes, 27–30
Io, 78
Italy, 30, 152
Izmit (Turkey), 197

Jamaica, 21–22
Japan
    earthquake prediction, 211, 214–15
    earthquakes in, 56–58, 70, 98, 128
    and plate tectonics, 8, 26
    tsunamis, 180
    Unzen volcano, 180
Johnston, Arch C., 89
JOIDES *Resolution*, 14
Jones, Lucy, 44
Jonesboro fault, 114
Jorullo volcano (Mexico), 149
Juan de Fuca plate, 151
junctions, triple, 20
Jupiter, moons of, 78

Kamchatka volcano (Russia), 154
Kanamori, Hiroo, 61
Kárník, Vít, 58
Kashmir (India and Pakistan), 186
Kauai (Hawaii), 160
Kilauea volcano (Hawaii), 153–54, *163*, 204
King, Geoffrey, 196
Kobe (Japan), 70, 98, 128
Krakatau volcano (Indonesia), 147, 150, 173

lahars, 152
Lake Roosevelt (Washington), 182
Lake Tahoe (California), 181
Lake Vostok (Antarctica), 144
lakes and tsunamis, 180–82
Lanark (Ontario), 119
Landers (California), 155, 195
landslides, 185–91
Latah Creek fault, 200
lateral spreading, *165*
Laurentide ice sheets, 104
lava flows, 148–49, 151, 152
Lawrenceville (Illinois), 110
Lay, Thorne, 174
lightning, 67
lights and earthquakes, 70–71
Lin, Cheng Horng, 76
liquefaction
    definition of, 40
    and earthquakes, 86, 89–90, 98, 113, 227
    lateral spreading, *165*
    and the Wabash Valley fault system, 107–8
Lisbon (Portugal), 194
Little Prairie (Missouri), 83–84, 86
Little Skull Mountain (Nevada), 167
local magnitude scale
    see Richter magnitude test scale
Loihi (Hawaii), 160
Loma Prieta (California), 35, 38–39, *55*, *165*, 209, 244
Long Island Sound (New York), 120
long-period earthquakes, 153–54
Long Valley caldera, 154
Los Angeles Basin, 30, 39
Los Angeles (California), 30, 62, 96–97
Love, A. E. H., 36
Love waves, 36
luminescence, 71
lunar earthquakes, 76–77

Machu Picchu (Peru), 189
Macquarie Island earthquake, 202
Madagascar, 8
magma, 7, 9, 14, 151, 153–56, 160–64
magnetism, 17
magnitude scales, 60–63
Maldives, 161
mantle, 7, 14–16, 229
mantle convection theory, 15–16, 18–19
March, 71
Mariana Arc, 29
Mariana Islands, 13, 218
Marianas trench, 12
Marked Tree (Arkansas), 91, 100
Mars, 78
Martinique, 150
Mascarene Plateau, 161
Massachusetts, 129–30, 240–41
Matsushiro (Japan), 70
Matuyama Reverse, 17
Mauna Loa volcano (Hawaii), 190
Mauritius Islands, 161
measuring earthquakes, 49–65
Meckering (Australia), 43, 138–39
Medvedev, Sergei, 58
Medvedev-Sponheuer-Karnik (MSK-64) scale, 57
Meeberrie (Australia), 138
Mercalli, Giuseppe, 52
Mercalli-Cancani-Sieberg scale (MCS scale), 53
Mercury, 78
MERMAID (Mobile Earthquake Recorder in Marine Areas by Independent Divers), 229
meteorite strikes, 77
Mexico, 148–49, 151
Mexico City (Mexico), 40
Michell, John, 64
micro-earthquakes, 203–4
microplates, 13, 20
microquakes, 153–54
Mid-Atlantic ridge, 8–10, 13, 16–17, 27, 151, 176–77
Mid-Indian ridge, 20
Middle America trench, 151
Mississippi River, 84, 92
Missouri, 45, 68, 81–104, 142
mistpouffers, 116
MM scales (Modified Mercalli intensity scales), 53–55

Mobile Earthquake Recorder in Marine Areas by Independent Divers (MERMAID), 229
Modified Mercalli intensity scales (MM scales), 53–55, 91–92
Moment Magnitude, 61–63
monitoring earthquakes, 164, 210, 228–29, 241–43
monogenetic field, 149
Montagua fault (Guatemala), 27
moon, 76–77
Mount Ranier (Washington), 151
Mount Saint Helens (Washington), 25, 154, 180–81
Mount Wrangell volcano (Alaska), 155
movies about earthquakes or volcanoes, 42, 148, 219
Mozambique, 140
Mt. Hood (Oregon), 202
mud-flows, 152

nanamkipoda, 130
National Earthquake Hazards Reduction Program, 237, 241–42
Natural and Statistical View (Drake), 92–94
Nazca plate, 12, 20, 27, 141
Netherlands, 136
Neumann, Frank, 53, 55
Nevada, 166–67, 218
New England, 129–32, 194, 240
New Hampshire, 131, 240
New Madrid fault zone, 31, 82, 95–100, 102–3, 235–36
New Madrid (Missouri), 45, 68, 81–104, 142, 194
New Orleans (steamboat), 84–85
New York, 4, 6, 103, 118–28, 133
New York City (New York), 121–28
New Zealand, 8, 27, 49
Nimitz freeway (California), 39
normal fault, 26
North American plate, 8, 24, 40, 122, 151, 200
North Anatolian fault, 196
North Carolina, 114–16
North Pole, 201
Northridge (California), 30, 62, 101, 217
Northwood, T. D., 234
nuclear waste, 166–67
Nutbush Creek fault, 115
Nuttli, Otto W., 90

ocean drilling, 14
Ohio, 85, 92–94, 101
Olympia (Washington), 28
Omori, Fusakichi, 44–45
Omori Seismic Intensity Scale, 56–57
Omori's law, 44–45
125th Street fault zone (New York), 122
Oregon, 202
Osmori, Fusakichi, 56
Ossippee (New Hampshire), 240
outcrops, 24

P-waves, 35–36, 61, 69, 101
Pacific Antarctic ridge, 20
Pacific Northwest, 28
Pacific Ocean, 27, 29, 170
Pacific plate
  and the Aleutian Islands, 151
  and earthquakes, 200, 218
  and elastic rebound theory, 24
  and plate tectonics, 8, 10, 12, 138
  and the Ring of Fire, 27
  and triple junctions, 20
Pacific Tsunami Warning Center, 170
Pakistan, 59, 186
Palo Alto (California), 3
Panabáj (Guatemala), 186
Pangaea, 15
Papua New Guinea, 180
Paradox Valley (Colorado), 221
Paricutín volcano (Mexico), 148–49
Parkfield (California), 195
Paso Robles (California), 52
Pennsylvania, 201
permafrost, 73–74
Peru, 189
Peru-Chile trench, 12
Philippine plate, 12, 27, 138
Philippines, 152, 185–86
piezoelectric effects, 70–71
Pinatubo volcano (Philippines), 152
Pinckney, Paul, 106
Piton de la Fournaise volcano (Reunion Island), 161
Piton Des Neiges volcano (Reunion Island), 161
plate boundaries, 5, 9–14, 11, 26
  see also faults
plate tectonics, 8–10, 12, 14–17, 22, 24, 26
Plattsburgh (New York), 4, 120
plumes, 156–57, 228

Plymouth (Massachusetts), 129
polar shifts, 17
Pompeii (Italy), 152
El Popocatépetl volcano (Mexico), 151
population growth, 216–17, 226
Port Royal (Jamaica), 21–22
ports, 227
Portugal, 194
potential disasters, 190–91
prediction, 43–44, 97–102, 197, 204, 208–15
preparedness, 96–97, 237–38, 247–63
Press, Frank, 223
primary waves, 35–36
Prince William Sound (Alaska), 62, 71
psychic predictions, 212
Puente Hills (California), 30
Pulido, Dionisio, 148–49

Quincy (Florida), 3

Rainer, J. H., 234
Ramapo fault system, 124
Rangeley Field (Colorado), 221
Ravenel (South Carolina), 113
Rayleigh waves, 36, 61
Red Sea, 10, 16
Redoubt volcano (Alaska), 154
Reelfoot Lake (Tennessee), 86
Reelfoot rift, 102–3
Reelfoot thrust fault, 31
Reid, Henry Fielding, 24
relay ramps, 30
remotely triggered earthquakes, 193–98
Reunion Island, 161
Revelations, 21
reverse fault, 26
Richter, Charles, 45, 60
Richter magnitude test scale, 45, 60–62
rift volcanoes, 151
Ring of Fire (Pacific Ocean), 27, 29, 141
risk assessment, 225–26
Ritter Island volcano (Papua New Guinea), 180
rock slides, 188
Rockaway (New York), 121
Rocks That Crackle and Sparkle and Glow (Freund), 67
Rocky Mountain Arsenal, 220–21
Rocky Mountains, 162
Roermond (Netherlands), 136
rogue earthquakes, 106
rogue tsunamis, 179

Rossi-Forel scale, 51–52
rupture surfaces
    see faults
Russia, 154

S-waves, 35–36, 69, 101
sag basins, 108
Saguenay (Canada), 70
Saint Francis Lake (Arkansas), 86
San Andreas Fault, 8, 13, 24, 26, 219
San Fernando (California), 231
San Francisco (California), 23–24, 38, 42–43, 62, 232
sand blows, 108–9, *111*
Sangre de Cristo fault, 221
Sannio (Italy), 30
Saudi Arabia, 10
Schiantarelli, Pompeo, 50
Schwartz, David, 232
Scotch Cap Coast Guard Station, 71, 168
Sea of Marmara (Turkey), 197
seafloor spreading, 17
seamount, 156
Seattle (Washington), *109*, 151
secondary waves, 35–36
Segundo (Colorado), 199–200
seiche, 117, 171
seismic creep, 46
seismic sea waves
    see tsunamis
seismic waves, 35–36, 84, 226
    see also Love waves; P-waves; Rayleigh waves; S-waves
seismograms, 63–65
seismograph stations, 216
seismographs, 63, 216
Seismological Society of America, 233
seismometers, 60, 63–65
Seneca guns, 116
Seward (Alaska), *156, 157*
Shanzi (China), 202
shear waves, 35–36
shindo scale, 58
shock waves, 35–36, 152
Sieberg, August, 53
Siever, Raymond, 223
silent earthquakes, 203–4
snowmelt, 202
soft sediment, 39–40
soil structure
    and earthquakes, 89, 94, 99–101, 113, 238
    and landslides, 188
    New England, 131
    New York City, 124–25
    and soft sediment, 39–40
    and the Wabash Valley fault system, 107–8
somoluminescence, 71
South Africa, 14
South America, 141–43
South American plate, 12, 27
South Carolina, 106, 110–14, *119, 123, 132*
Southern Ocean, 171
Spirit Lake (Washington), 180–81
Spokane (Washington), 68, 199–200
Sponheuer, Wilhelm, 58
Sri Lanka, 172–73
statistics, 209–11, 224–25, 230
Steele (Missouri), 84
Stein, Ross, 195
stratovolcanoes, 150
strike-slip faults, 26
Structural Engineers Association of California, 233
structures
    see buildings and structures
Strutt, John William, 36
subduction zones, 12, 26, 30, 141, 151, 154, 171
Sumatra-Andaman earthquake, 41, 59, 155, *160*, 174–75, 201, 225
Sumatra-Andaman Great Earthquake Research Initiative, 175
Summerville (South Carolina), 111–12
Sunda fault and trench, 171–72
surface-waves, 61, 174
Sutherland, Alex H., 33
swarms, 68, 153–54, 175, 199–200
Switzerland, 136
syzygy, 212

Taipei (Taiwan), 76
Taiwan, 76, 135
Tambora volcano (Indonesia), 150
Tangshan (China), 59, 210, 214, 223
tectonic plates, 8–10
Tennessee, 31, 86, 237–39
Thailand, 172
thermal earthquakes, 77
thermal plumes, 156–57
thrust fault, 26
Tibetan Plateau, 8–9, 12
tidal forces, 77

tidal waves
    see tsunamis
Timiskaming (Canada), 133
Tokat (Turkey), 196
tomographic images, 164
Tonga and the Ring of Fire, 27
Trans-Mexican volcanic belt, 149
transform boundaries, 9, 13
transform fault zone, 44
transverse motion waves, 36
triple junctions, 20, 41
tsunamis, 166–84
    in Alaska, 71, *156, 157*, 168
    and asteroid or comet strikes, 182–84
    education about, 181–82
    and faults, 42
    after Grand Banks earthquake, 133
    Hawaii, 71, 167–69, 171, *190*
    intensity scales, 53
    in Japan, 180
    in lakes, 180–82
    and the Mississippi River, 84
    Port Royal, 22
    rogues, 179
    Spirit Lake (Washington), 180–81
    Sumatra-Andaman earthquake, 41, 59, 155, *160*, 174–75
    in the United States, 176–79
    warning systems, 170, 176
turbidity currents, 134
Turkey, 59, 196–97
types of earthquakes, 9

U. S. Bureau of Reclamation, 221
undersea cables, 134–35
underwater slides, 191
underwater volcanic mountains, 156
Uniform Building Code, 235
United Kingdom, 136
United States, 176–79, 225–26, 241–42
    see also specific locations
United States Geological Survey, 63, 221, 232, 237–38, 248
Unzen volcano (Japan), 180
U.S. Virgin Islands, 179
USGS
    see United States Geological Survey

van der Vink, G., 207
VAN method of earthquake prediction, 211
velocity, 35
Venezuela, 141–42
Venus, 78
View to a Kill, A (movie), 219
volcanic ash, 149, 151, 152
volcanic tremor, 153–54
volcanoes, 148–58
    and earthquakes, 152–55
    rift volcanoes, 151
    stratovolcanoes, 150
    and water, 155
    see also specific volcanoes and locations

Wabash Valley fault system, 107–10
Ward, Steven, 183
Warm Springs (Arkansas), 217
Washington
    earthquakes in, 28, 68, 199–200
    Grand Coulee Dam, 182
    Lake Roosevelt, 182
    Mount Ranier, 151
    Mount Saint Helens, *25*, 154, 180–81
    Spirit Lake, 180–81
water, effect of, 155, 200–202
weather and tsunamis, 169
weather and volcanoes, 72–75
websites, 249, 253–61
Wegener, Alfred, 15
well water, 200–202, 220–21
Wilson, J. Tuzo, 155–56
Wood, Harry, 53, 55
Wood-Anderson torsion seismometer, 60
Woodstock fault, 114
Woodstock (South Carolina), 113

Yellowstone National Park, 161–64, *177*
Yosemite National Park, 189
Yucca Mountain (Nevada), 166–67

"zebra stripes," 17
Zoback, Mark, 103